T0305773

# Event History Analysis
# with R

# Event History Analysis with R

## Second Edition

Göran Broström

CRC Press
Taylor & Francis Group
Boca Raton  London  New York

CRC Press is an imprint of the
Taylor & Francis Group, an **informa** business

A CHAPMAN & HALL BOOK

First edition published 2012
by CRC Press
6000 Broken Sound Parkway NW, Suite 300, Boca Raton, FL 33487-2742

and by CRC Press
2 Park Square, Milton Park, Abingdon, Oxon, OX14 4RN

© 2022 Taylor & Francis Group, LLC

First edition published by CRC Press 2012

CRC Press is an imprint of Taylor & Francis Group, LLC

ISBN: 978-1-138-58771-7 (hbk)
ISBN: 978-1-032-12320-2 (pbk)
ISBN: 978-0-429-50376-4 (ebk)

DOI: 10.1201/9780429503764

Typeset in Latin Modern font
by KnowledgeWorks Global Ltd.

# Contents

# List of Tables

# List of Figures

# *Preface*

The first edition of this book was published in 2012. Since then the field of event history and survival analysis has developed rapidly, both in terms of available scientific data and of software development for analyzing data, not the least in the **R** environment. We have also seen the public need of analyzing what is going on around the world, and the present (2021) state of the COVID-19 pandemic is a striking example of that.

So the time for a second edition of the book is now. The basic chapters on Cox regression and proportional hazards modeling are much the same as in the first edition, but they have been updated. There are two new chapters, Chapter 4, *Explanatory Variables and Regression*, and Chapter 7, *Register-Based Survival Data Models*. Compared to the first edition, chapters have been reordered so that the logical flow is clearer. On the other hand, the appendices C and D are somewhat shorter now because there are a lot of excellent sources on-line today covering their topic, which anyway is a little bit off here.

Since the publication of the first edition, focus has gradually shifted toward the analysis of large and huge data sets, where Cox regression favorably can be replaced by parametric proportional hazards models with piecewise constant baseline hazard functions. With huge data sets with excessively many events, this way of tabulating data leads to a significant reduction of necessary efforts in producing reliable results with the same precision and power a full analysis would have had. This result relies on reduction by the mathematical *sufficiency principle*, all irrelevant (so defined by the model) noise is eliminated. A word of warning is that this noise (*residuals*) may be relevant in *model evaluation*.

A second point of importance is how to present results from a regression analysis, and especially how to present estimated *p*-values. Two important issues in this area are (i) present only *relevant* ones, and (ii) make sure they are of the right kind, that is, *likelihood ratio based*. This important topic is addressed in the book, and supported by new summary functions in the **R** package eha (Broström, 2021).

The writing of this book has been done in parallel with the development of the **R** package eha. Almost all the data sets used in the examples in this book are available in eha, so you can easily play around with them on your own. Some data sets will also be published on the home page of the package, http://ehar.se/r/eha/.

I had, as usual, invaluable support from the publisher, CRC Press, Ltd. I especially want to thank Vaishali Singh and Rob Calver for their interest in my work and their encouragement in the project.

The first edition of the book was written in *LaTeX* with support of the **R** package Sweave, but with the second edition we decided to do the writing in Rmarkdown using the **R** packages bookdown and knitr. The reason was mainly that it allowed for the production of output in both *HTML* (for the website) and *PDF* (the printed book). I am indepted to Yihui Xie for his important work in this area, which made this approach possible.

Adding to the list of people who gave valuable input to the First Edition of the book: Kristian Hindberg and Glenn Sandström have contributed with suggestions that have improved the text in this Second Edition. Many thanks goes to Elisabeth Engberg, director of the Centre for Demographic and Ageing Research (CEDAR), Umeå University, for letting me use the facilities that made this work possible.

Göran Broström
professor emeritus
CEDAR, Umeå University
Umeå, Sweden
April 2021

# *Preface to the First Edition*

This book is about the analysis of event history and survival data, with special emphasis on how to do it in the statistical computing environment R (R Core Team, 2021). The main applications in mind are demography and epidemiology, but also other applications, where durations are of primary interest, fit in to this framework. Ideally, the reader has a first course in statistics, but some necessary basics may be found in Appendix A. The basis for this book is courses in event history and survival analysis that I have given and developed over the years since the mid-eighties, and the development of software for the analysis of this kind of data. This development has during the last ten to fifteen years taken place in the environment of R in the package **eha** (Broström, 2021). There are already several good text books in the field of survival and event history analysis on the market, Aalen et al. (2008), Andersen et al. (1993), Cox and Oakes (1984), Hougaard (2000), Kalbfleisch and Prentice (2002), and Lawless (2003) to mention a few. Some of these are already classical, but they all are aimed at an audience with solid mathematical and statistical background. On the other hand, the Parmar and Machin (1995), Allison (1984), Allison (1995), Klein and Moeschberger (2003), Collett (2003), and Elandt-Johnson and Johnson (1999) books are all more basic but lack the special treatment demographic applications needs today, and also the connection to **R**.

In the late seventies, large databases with individual life histories began to appear. One of the absolutely first was *The Demographic Data Base* (DDB) at Umeå University. I started working there as a researcher in 1980, and at that time the data base contained individual data for seven Swedish parishes scattered all over the country. Statistical software for Cox regression (Cox, 1972) did not exist at that time, so we began a development, with the handling of large data sets in mind, of Fortran programs for analyzing censored survival data. The starting point was Appendix 3 of Kalbfleisch and Prentice (1980). It contained "Fortran programs for the proportional hazards model."

The beginning of the eighties was also important because then the first text books on survival analysis began to appear. Of special importance (for me) was the books by Kalbfleisch and Prentice (1980), and, four years later, Cox and Oakes (1984).

The time period covered by the DDB data is approximately the 19th century. Today the geographical content is expanded to cover four large regions, with more than 60 parishes in total. This book is closely tied to the *R Environment for Statistics and Computing* (R Core Team, 2021). This fact has one specific advantage: Software and this book can be totally synchronized, not only by adapting the book to the software, but also vice versa. This is possible because R and its packages are *open source*, and one of the survival analysis packages (Broström, 2021) in R and this book have the author in common. The `eha` package contains some research results not found in other software (Broström and Lindkvist, 2008; Broström, 2002, 1987). However, it is important to emphasize that the real work horse in survival analysis in R is the recommended package `survival` (Therneau, 2021).

The mathematical and statistical theory underlying the content of this book is exposed to a minimum; the main target audience is social science researchers and students who are interested in studying demographically oriented research questions. However, the apparent limitation to demographic applications in this book is not really a limitation, because the methods described here are equally useful whenever durations are of primary interest, for instance in epidemiology and econometrics.

In Chapter 1 event history and survival data are exemplified, and the specific problems data of this kind poses on the statistical analysis. Of special importance are the concepts of censoring and truncation, or in other words, incomplete observations. The dynamic nature of this kind of data is emphasized. The data sets used throughout the book are presented here for the first time.

How to analyze homogeneous data is discussed in Chapter 2. The concept of a survival distribution is introduced, including the hazard function, which is the fundamental concept in survival analysis. The functions that can be derived from the hazard function, the survival and the cumulative hazard functions, are introduced. Then the methods for estimating the fundamental functions nonparametrically are introduced, most important the Kaplan-Meier and Nelson-Aalen estimators. By "nonparametric" is meant that in the estimation procedures, no restrictions are put on the class of allowed distributions, except that it must consist of distributions on the positive real line; a life length cannot be negative. Some effort is put into giving an understanding of the intuitive reasoning behind the estimators, and the goal is to show that the ideas are really simple and elementary.

Cox regression is introduced in Chapter 3 and expanded on in Chapter 5. It is based on the property of proportional hazards, which makes it possible to analyze the effects of covariates on survival without specifying a family of survival distributions. Cox regression is in this sense a non-parametric method, but correct is perhaps to call it semi-parametric, because the proportionality constant is parametrically determined. The introduction to Cox regression goes via the log-rank test, which can be regarded as a special case of Cox regression. A fairly detailed discussion of different kinds of covariates are given, together with an explanation of how to interpret regression parameters connected to these covariates. Discrete time versions of the proportional hazards assumption is introduced.

In Chapter 4, a break from Cox regression is taken, and Poisson regression is introduced. The real purpose of this, however, is to show the equivalence (in some sense) between Poisson and Cox regression. For tabular data, the proportional hazards model can be fitted via Poisson regression. This is also true for the continuous time piecewise constant hazards model, which is explored in more detail in Chapter 6.

The Cox regression thread is taken up again in Chapter 5. Time-varying and so-called communal covariates are introduced, and it is shown how to prepare a data set for these possibilities. The important distinction between internal and external covariates is discussed. Stratification as a tool to handle non-proportionality is introduced. How to check model assumptions, in particular the proportional hazards one, is discussed, and finally some less common features, like sampling of risk sets and fixed study period, are introduced.

Chapter 6 introduces fully parametric survival models. They come in three flavors, proportional hazards, accelerated failure time, and discrete time models. The parametric proportional hazards models have the advantages compared to Cox regression that the estimation of the baseline hazard function comes for free. This is also part of the drawback with parametric models; they are nice to work with but often too rigid to fit data from a complex reality. It is for instance not possible to find simple parametric descriptions of human mortality over the full life span, it would require a U-shaped hazard function, and no standard survival distribution has that property. However, when studying shorter segments of human life span, very good fits may be achieved with parametric models. So is for instance old age mortality, say above the age of 60, possible to model accurately with the Gompertz distribution. Another important general feature of parametric models is that they convey a a simple and clear message of the properties of a distribution or relation, thus easily adds to the accumulated knowledge about the phenomenon it is related to.

The accelerated failure time (AFT) model is an alternative to the proportional hazards (PH) one. While in the PH model, relative effects are assumed to remain constant over time, the AFT model allows for effects to shrink toward zero with age. This is sometimes a more realistic scenario, for instance in a medical application, where a specific treatment may have an instant, but transient effect.

With modern register data, exact event times are often not available. Instead data are grouped in one-year intervals. This is generally not because exact information is missing at the government agencies, but a result of integrity considerations. With access to birth date information, a person may be easy to identify. Anyway, as a result age is only possible to measure in (completed) years, and heavily tied data sets will be the result. That opens up for the use of discrete-time models, similar to logistic and Poisson regression. In fact, as is shown in Chapter 4, there is a close relation between Cox regression and binary and count data regression models.

Finally, in the framework of Chapter 6 and parametric models, it is important to mention the piecewise constant hazard (PCH) model. It constitutes an excellent compromise between the non-parametric Cox regression and the fully parametric models discussed above. Formally, it is a fully parametric model, but it is easy to adapt to given data by changing the number of cut points (and by that the number of unknown parameters). The PCH model is especially useful in demographic and epidemiological applications where there is a huge amount of data available, often in tabular form.

Chapters 1–6 are suitable for an introductory course in survival analysis, with a focus on Cox regression and independent observations. In the remaining chapters, various extensions to the basic model are discussed.

In Chapter 7, a simple dependence structure is introduced, the so-called shared frailty model. Here it is assumed that data are grouped into clusters or litters, and that individuals within a cluster share a common risk. In demographic applications the clusters are biological families, people from the same geographical area, and so on. This pattern creates dependency structures in the data, which are necessary to consider in order to avoid biased results.

Competing risks models are introduced in Chapter 8. Here the simple survival model is extended to the inclusion of failures of several types. In a mortality study it may be deaths in different causes. The common feature in these situations is that to each individual, many events may occur, but at most one will occur. The events are competing with each other, and at the end there is only one winner. It turns out that in these situations the concept of cause-specific hazards is meaningful, while the discussion of cause-specific survival functions is problematic.

During the last few decades, causality has become a hot topic in statistics. In Chapter 9 a review of parts relevant to event history analysis are given. The concept of matching is emphasized. It is also an important technique of its own, not necessarily tied to causal inference. There are four appendices. They contain material that is not necessary to read in order to be able to follow the core of the book, given proper background knowledge. Browsing through Appendix A is recommended to all readers, at least so that common statistical terminology is agreed upon. Appendix B contains a description of relevant statistical distributions in R, and also a presentation of the modeling behind the parametric models in the R package **eha**. The latter part is not necessary for the understanding of the rest of the book. For readers new to R, Appendix C is a good starting point, but it is recommended to complement it with one of many introductory text books on R or on-line sources. Consult the home of R[1] and search under *Documentation*. Appendix C also contains a separate section with a selection of useful functions from the **eha** and the **survival** packages. This is of course recommended reading for everyone. It is also valuable as a reference library to functions used in the examples of the book. They are not always documented in the text.

Finally, Appendix D contains a short description of the most important packages for event history analysis in R.

As a general reading instruction, the following is recommended. Install the latest version of R from CRAN[2] and replicate the examples in the book by yourself. You need to install the package **eha** and load it. All the data in the examples are then available on line in R.

Many thanks go to people who have read early and not so early versions of the book; they include Tommy Bengtsson, Kristina

---

[1] https://www.r-project.org

[2] https://cran.r-project.org

Broström, Bendix Carstensen, Renzo Derosas, Sören Edvinsson, Kajsa Fröjd, Ingrid Svensson, and students at the Departments of Statistics and Mathematical Statistics at the University of Umeå. Many errors have been spotted and improvements suggested, and the remaining errors and weaknesses are solely my responsibility.

I also had invaluable support from the publisher, CRC Press, Ltd. I especially want to thank Sarah Morris and Rob Calver for their interest in my work and their thrust in the project.

Of course, without the excellent work of the R community, especially the R Core Team, a title like the present one had been impossible. Especially I want to thank Terry Therneau for his inspiring work on the **survival** package; the **eha** package depends strongly on it. Also, Friedrich Leisch, author of the **Sweave** package, deserves many thanks. This book was entirely written with the aid of the package **Sweave** and the typesetting system *LaTeX*, wonderful companions.

Finally, I want to thank Tommy Bengtsson, director of the *Centre for Economic Demography*, Lund University, and Anders Brändström, director of the Demographic Data Base at Umeå University, for kindly letting me use real data in this book and in the R package **eha**. It greatly enhances the value of the illustrations and examples.

Göran Broström

Umeå, Sweden

# *Author*

Göran Broström is a professor emeritus of Statistics at the Centre for Demographic and Ageing Research, Umeå University, Sweden. He has a PhD in mathematical statistics from Umeå University (1979). He is the author of two R packages, eha and glmmML, available on CRAN

# 1

## Event History and Survival Data

The thing that characterizes event history and survival data the most is its *dynamic* nature. Individuals are followed over time, and during that course, the timings of events of interest are noted. Naturally, things may happen that makes it necessary to interrupt an individual follow-up, such as the individual suddenly disappearing for some reason. With classical statistical tools, such as linear regression, these observations are difficult or impossible to handle in the analysis. The methods discussed in this book aim among other things, at solving such problems.

In this introductory chapter, the special techniques that constitute survival and event history analysis are motivated. The concepts of *right censoring* and *left truncation* are defined and discussed. The data sets used throughout the book are also presented in this chapter.

The environment R is freely (under a *GPL license*) available for download from CRAN[1]. There you will find precompiled versions for *Linux*, *Mac OS*, and *Microsoft Windows*, as well as the full *source code*, which is open. See *Appendix C* for more information.

## 1.1  Survival Data

Survival data (survival times) constitute the simplest form of event history data. A survival time is defined as the time it takes for an event to occur, measured from a well-defined start event. Thus,

---

[1]https://cran.r-project.org

DOI: 10.1201/9780429503764-1

there are three basic elements which must be well defined: a time origin, a scale for measuring time, and an event. The response in a statistical analysis of such data is the exact time elapsed from the time origin to the time at which the event occurs. The challenge, which motivates special methods, is that in most applications, this duration is often not possible to observe exactly.

As an introduction to the research questions that are suitable for handling with event history and survival analysis, let us look at a data set found in the **eha** package (Broström, 2021) in R (R Core Team, 2021).

**Example 1.1** (Old age mortality)

The data set *oldmort* in **eha** contains survival data from the parish Sundsvall in the mid-east of 19th century Sweden. The name *oldmort* is an acronym for *old age mortality.* The source is digitized information from historical parish registers, church books. More information about this can be found at the web page of the *Centre for Demographic and Ageing Research* at Umeå University (CEDAR[2]).

The sampling was done as follows: Every person who was present and alive and 60 years of age or above anytime between January 1, 1860 and December 31, 1879 was followed from the entrance age (for most people that would be 60) until the age when last seen, determined by death, out-migration, or surviving until December 31, 1879. Those born during the 18th century would enter observation at an age above 60, given that they lived long enough, that is at least until January 1, 1860.

Two types of finishing the observation of a person are distinguished: Either it is by *death* or it is by something else, out-migration or end of study period. In the first case we say that the event of interest has occurred, in the second case not.

After installing the **eha** package and starting an R session (see Appendix C), the data set is loaded by loading **eha** as follows.

---

[2]https://www.umu.se/cedar

```
library(eha)
```

Let us look at the first few lines of *oldmort* (slightly manipulated). It is conveniently done with the aid of the R function **head**:

```
head(oldmort, 3)
```

```
  id enter   exit event birthdate m.id f.id    sex       civ ses.50
1  1 94.510 95.813  TRUE  1765.490   NA   NA female     widow unknown
2  2 94.266 95.756  TRUE  1765.734   NA   NA female unmarried unknown
3  3 91.093 91.947  TRUE  1768.907   NA   NA female     widow unknown
  birthplace imr.birth   region
1     remote  22.20000    rural
2     parish  17.71845 industry
3     parish  12.70903    rural
```

Note that the first printed column, without a name, contains the *row names*. If no row names are explicitly given, they are set to the *row numbers*. If you don't want row names printed, you have to use the more cumbersome

```
print(oldmort[1:3, ], row.names = FALSE)
```

```
 id enter   exit event birthdate m.id f.id    sex       civ ses.50
  1 94.510 95.813  TRUE  1765.490   NA   NA female     widow unknown
  2 94.266 95.756  TRUE  1765.734   NA   NA female unmarried unknown
  3 91.093 91.947  TRUE  1768.907   NA   NA female     widow unknown
 birthplace imr.birth   region
     remote  22.20000    rural
     parish  17.71845 industry
     parish  12.70903    rural
```

There are several packages for nice printing of tables available on CRAN. In this book, written with the aid of the **R** packages bookdown (Xie, 2016, 2020), knitr (Xie, 2015, 2021), and kableExtra (Zhu, 2021), outputs of tables and regression results are first given "asis", that is, as you will see them on the computer screen, but later, depending on circumstances, nice printing is preferred. Code like the following is often used for regression results:

```
fit.out <- function(fit, caption, label){
    if (knitr::is_latex_output()){ # PDF
        if (!missing(label)){
            label <- paste0("tab:", label)
        }
        dr <- drop1(fit, test = "Chisq")
        ltx(fit, dr = dr, caption = caption, label = label)
    }else{ # HTML
        xx <- regtable(summary(fit), digits = 4)
        nn <- ncol(xx)
        rr <- c("Max Log", "Likelihood", "",
            round(fit$loglik[2], 1), rep("", nn - 4))
        xx <- rbind(xx, rr)
        kbl(xx, booktabs = TRUE, caption = caption) %>%
            kable_styling(font_size = 12,
                full_width = FALSE)
    }
}
```

Essentially, if output is a *pdf* document (the printed book), the function ltx (from the package **eha**) is used, and if output is in HTML format, the functionality of the packages **knitr** and **kableExtra** is used.

For rectangular tables (for instance, data), the code is simpler, we can use **kableExtra** for both HTML and pdf output:

**TABLE 1.1** Old age mortality, 19th century Sundsvall, Sweden.

| enter | exit | event | birthdate | sex | civ | region |
|-------|------|-------|-----------|-----|-----|--------|
| 65.201 | 70.712 | FALSE | 1809-04-16 | female | widow | industry |
| 72.853 | 73.699 | FALSE | 1806-04-21 | female | widow | town |
| 70.947 | 77.020 | TRUE | 1795-09-16 | female | widow | industry |
| 60.000 | 60.987 | FALSE | 1803-07-04 | male | married | industry |
| 60.000 | 65.827 | FALSE | 1814-03-05 | male | married | rural |
| 60.000 | 74.263 | FALSE | 1805-09-27 | female | married | industry |
| 66.734 | 68.086 | FALSE | 1805-05-28 | male | widow | rural |
| 60.000 | 69.096 | FALSE | 1810-11-27 | female | widow | rural |
| 63.153 | 66.994 | TRUE | 1796-11-05 | female | widow | rural |
| 60.943 | 64.431 | FALSE | 1799-01-21 | female | married | rural |

```
tbl <- function(tt, caption = "", fs = 11){
    library(kableExtra)
    kbl(tt, caption = caption, booktabs = TRUE, row.names = FALSE) %>%
        kable_styling(full_width = FALSE, font_size = fs,
                      position = "center")
}
```

The result (with selected variables and a random sample of records) is shown in Table 1.1.

The variables in *oldmort* have the following definitions and interpretations:

- **id** A unique id number for each individual.
- **enter, exit** The start age and stop age for this record (spell). For instance, in row No. 1, individual No. 1 enters under observation at age 94.51 and exits at age 95.81. Age is calculated as the number of days elapsed since birth and this number is then divided by 365.25 to get age in years. The denominator is the average length of a year, taking into account that every fourth year is 366 days long. The first individual was born on June 27, 1765, and so almost 95 years of age when the study started. Suppose that this woman had died at age 94; then she had not

been in our study at all. This property of our sampling procedure introduces a phenomenon known as *length-biased sampling*. That is, of those born in the 18th century, only those who live well beyond 60 will be included. This bias must be compensated for in the analysis, and it is accomplished by conditioning on the fact that these persons were alive at January 1, 1860. This technique is called *left truncation*.

- **event** A logical variable (taking values *TRUE* or *FALSE*) indicating if the exit is a death (*TRUE*) or not (*FALSE*). For our first individual, the value is *TRUE*, indicating that she died at the age of 95.81 years. Another possible coding is to use *event* = *1* for a death and *event* = *0* otherwise.

- **birthdate** The birth date expressed as the time (in years) elapsed since January 1, year 0 (which by the way does not exist). For instance, the (pseudo) date 1765.490 is really June 27, 1765. The fraction 0.490 is the fraction of the year 1765 that elapsed until the birth of individual No. 1. However, here it is printed as a real date.

- **m.id** *Mother's id.* It is unknown for all the individuals listed above. That is the symbol *NA*, which stands for *Not Available*. The oldest people in the data set typically have no links to parents.

- **f.id** *Father's id.* See *m.id*.

- **sex** A categorical variable with the levels *female* and *male*.

- **civ** Civil status. A categorical variable with three levels; *unmarried, married*, and *widow(er)*.

- **ses.50** Socio-economic status (SES) at age 50. Based on occupation information. There is a large proportion of *NA* (missing values) in this variable. This is quite natural, because this variable was of secondary interest to the record holder (the priest in the parish). The occupation is only noted in connection to a vital event in the family (such as a death, birth, marriage, or in- or out-migration). For those who were above 50 at the start of the period there is no information on SES at 50.

- **birthplace** A categorical variable with two categories, *parish* and *remote*, representing born in parish and born outside parish, respectively.

- **imr.birth** A rather specific variable. It measures the infant mortality rate in the birth parish at the time of birth (per cent).
- **region** Present geographical area of residence. The parishes in the region are grouped into three regions, *Sundsvall town*, *rural*, and *industry*. The industry is the sawmill one, which grew rapidly in this area during the late part of the 19th century. The Sundsvall area was in fact one of the largest sawmill areas in Europe at this time.

Of special interest is the triple (enter, exit, event), because it represents the *response variable*, or what can be seen of it. More specifically, the sampling frame is all persons observed to be alive and above 60 years of age between January 1, 1860 and December 31, 1879. The start event for these individuals is their 60th anniversary and the stop event is death. Clearly, many individuals in the data set did not die before January 1, 1880, so for them we do not know the full duration between the start and stop events; such individuals are said to be *right censored* (the exact meaning of which will be given soon). The third component in the survival object (enter, exit, event), event, is a *logical* variable taking the value TRUE if exit is the true duration (the interval ends with a death) and FALSE if the individual is still alive at the duration "last seen".

Individuals aged 60 or above between January 1, 1860 and December 31, 1879 are included in the study. Those who are above 60 at this start date are included only if they did not die between the age of 60 and the age at January 1, 1860. If this is not taken into account, a bias in the estimation of mortality will result. The proper way of dealing with this problem is to use *left truncation*, which is indicated by the variable enter. If we look at the first rows of *oldmort* we see that the enter variable is very large; it is the age for each individual at January 1, 1860. You can add enter and *birthdate* for the first six individuals to see that:

```
oldmort$enter[1:6] + oldmort$birthdate[1:6]
```

```
[1] 1860 1860 1860 1860 1860 1860
```

The statistical implication (description) of left truncation is that its presence forces the analysis to be *conditional* on survival up to the age `enter`.

A final important note: In order to get the actual duration at exit, we must subtract 60 from the value of `exit`. When we actually perform a survival analysis in R, we should subtract 60 from both `enter` and `exit` before we begin. It is not absolutely necessary in the case of Cox regression, because of the flexibility of the baseline hazard in the model (it is in fact left unspecified!). However, for parametric models, it may be important in order to avoid dealing with truncated *distributions*.

Now let us think of the research questions that could be answered by analyzing this data set. Since the data contain individual information on the length of life after 60, it is quite natural to study what determines a long life and what are the conditions that are negatively correlated with long life. Obvious questions are: (i) Do women live longer than men? (Yes), (ii) Is it advantageous for a long life to be married? (Yes), (iii) Does socio-economic status play any role for a long life? (Don't know), and (iv) Does place of birth have any impact on a long life, and if so, is it different for women and men?  □

The answers to these, and other, questions will be given later. The methods in later chapters of the book are all illustrated on a few core examples. They are all presented at first time in this chapter.

The data set *oldmort* contained only two states, referred to as *Alive* and *Dead*, and one possible transition, from Alive to Dead, see Figure 1.1.

The ultimate study object in survival analysis is the time it takes from entering state *Alive* (e.g., becoming 60 years of age) until

**FIGURE 1.1** Survival data.

entering state *dead* (e.g., death). This time interval is defined by the exact time of two events, which we may call *birth* and *death*, although in practice these two events may be almost any kind of events. Economists, for instance, are interested in the duration of out-of-work spells, where "birth" refers to the event of losing the job, and "death" refers to the event of getting a job. In a clinical trial regarding treatment of cancer, the starting event time may be time of operation, and the final event time is time of relapse (if any).

## 1.2   Right Censoring

When an individual is lost to follow-up, we say that she is *right censored*, see Figure 1.2.

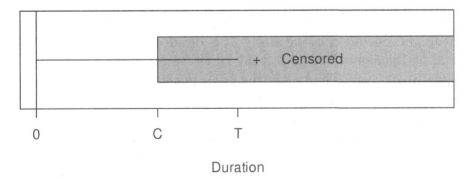

**FIGURE 1.2** Right censoring.

As indicated in Figure 1.2, the true age at death is $T$, but due to right censoring, the only information available is that death age $T$ is larger than $C$. The number $C$ is the age at which this individual was last seen. In ordinary, classical regression analysis, such data are difficult, if not impossible, to handle. Discarding such information may introduce bias. The modern theory of survival analysis offers simple ways to deal with right censored data.

A natural question to ask is: If there is right censoring, there should be something called *left* censoring, and if so, what is it? The answer to that is that yes, *left censoring* refers to a situation where the only thing known about a death age is that it is *less than* a certain value $C$. Note carefully that this is different from *left truncation*, see the next section 1.3.

---

## 1.3 Left Truncation

The concept of *left truncation*, or *delayed entry*, is well illustrated by the data set *oldmort* that was discussed in detail in Example 1.1. Please note the difference compared to *left censoring*. Unfortunately, you may still see scientific articles where these two concepts are mixed up.

It is illustrative to think of the construction of the data set *oldmort* as a statistical follow-up study, starting on January 1, 1860. At that day, all persons present in the parish and 60 years of age *or above*, are included in the study. It is decided that the study will end at December 31, 1879, that is, the study period (follow-up time) is 20 years. The interesting event in this study is *death*. This means that the start event is the sixtieth anniversary of birth and the final event is death. Due to the calendar time constraints (and migration), all individuals will not be observed to die (especially those who live long), and moreover, some individuals will enter the study after the "starting" event, the sixtieth anniversary. A person who enter late, say he is 65 on January 1, 1860, had not been included had he died at age 63 (say). Therefore, in the analysis, we must *condition* on the fact that he was alive at 65. Another way of saying this is to say that this observation is *left truncated* at age 65.

People being too young at the start date will be included from the day they reach 60, if that happens before the closing date, December 31, 1879. They are not left truncated, but will have a higher and higher probability of being right censored, the later they enter the study.

## 1.4 Time Scales

In demographic applications *age* is often a natural time scale, that is, time is measured from birth. In the old age data just discussed, time was measured from age 60 instead. In a case like this, where there is a common "late start age", it doesn't matter much, but in other situations it does. Imagine for instance that interest lies in studying the time it takes for a woman to give birth to her first child after marriage. The natural way of measuring time is to start the clock at the day of marriage, but a possible (but not necessarily recommended!) alternative is to start the clock at some (small) common age of the women, for instance at birth. This would give left truncated (at marriage) observations, since women were sampled at marriage. There are two clocks ticking, and you have to make a choice.

Generally, it is important to realize that there often are alternatives, and that the result of an analysis may depend strongly on the choice made.

### 1.4.1 The Lexis diagram

Two time scales are nearly always present in demographic research: Age (or duration) and calendar time. For instance, an investigation of mortality may be limited in these two directions. In Figure 1.3 this is illustrated for a study of old age mortality during the years

1829 and 1895. "Old age mortality" is defined as mortality from age 50 and onward to age 100. The Lexis diagram is a way of showing the interplay between the two time scales and (human) life lines. Age moves vertically and calendar time horizontally, which will imply that individual lives will move diagonally, from birth to death, from south-west to north-east, in the Lexis diagram. In our example study, we are only interested in the part of the life lines that appear inside the small rectangle.

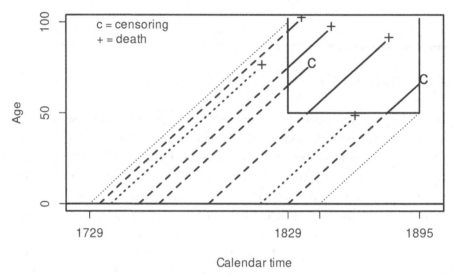

**FIGURE 1.3** Lexis diagram: time period 1829–1894 and ages 50–100.

Assume that the data set at hand is saved in the text file 'lex.dat'. Note that this data set is not part of **eha**; it is only used here for the illustration of the Lexis diagram.

```
lex <- read.table("Data/lex.dat", header = TRUE)
lex
```

```
  id enter   exit event birthdate   sex
1  1     0 98.314     1  1735.333  male
2  2     0 87.788     1  1750.033  male
```

```
3  3    0 71.233    0  1760.003 female
4  4    0 87.965    1  1799.492   male
5  5    0 82.338    1  1829.003 female
6  6    0 45.873    1  1815.329 female
7  7    0 74.112    1  1740.513 female
```

How do we restrict the data to fit into the rectangle given by the Lexis diagram in Figure 1.3? With the two functions **age.window** and **cal.window** in eha, it is easy. The former fixes the 'age cut' while the latter makes the 'calendar time cut'.

The age cut:

```
lex <- age.window(lex, c(50, 100))
lex
```

```
   id enter    exit event birthdate    sex
1  1    50 98.314    1  1735.333   male
2  2    50 87.788    1  1750.033   male
3  3    50 71.233    0  1760.003 female
4  4    50 87.965    1  1799.492   male
5  5    50 82.338    1  1829.003 female
7  7    50 74.112    1  1740.513 female
```

Note that individual No. 6 dropped out completely because she died too young. Then the calendar time cut:

```
lex <- cal.window(lex, c(1829, 1895))
lex
```

```
   id  enter   exit event birthdate    sex
1  1 93.667 98.314    1  1735.333   male
2  2 78.967 87.788    1  1750.033   male
3  3 68.997 71.233    0  1760.003 female
4  4 50.000 87.965    1  1799.492   male
5  5 50.000 65.997    0  1829.003 female
```

and here individual No. 7 disappeared because she died before January 1, 1829. Her death date is her birth date plus her age at death, $1740.513 + 74.112 = 1814.625$, or August 17, 1814.

---

### 1.5 Event History Data

Event history data arise, as the name suggests, by following subjects over time and making notes about what happens and when. Usually the interest is concentrated to a few specific kinds of events. The main application in this book is demography and epidemiology, hence events of primary interest are *births*, *deaths*, *marriages* and *migration*.

**Example 1.2** (Marital fertility in 19th century Sweden)

As a rather complex example, let us look at *marital fertility* in 19th century Sweden, see Figure 1.4.

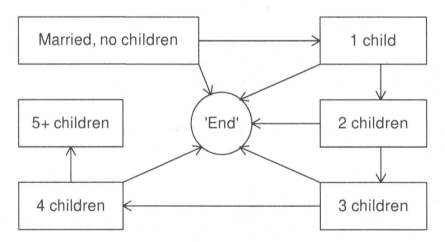

**FIGURE 1.4** Marital fertility.

In a marital fertility study, women are typically followed over time from the time of their marriage until the time the marriage is dissolved or her fertility period is over, say at age 50, whichever comes first. The marriage dissolution may be due to the death of the woman or of her husband, or it may be due to a divorce. If the study is limited to a given geographical area, women may get lost to follow-up due to out-migration. This event gives rise to a *right-censored* observation.

During the follow-up, the exact timings of child births are recorded. Interest in the analysis may lie in investigating which factors, if any, that affect the length of birth intervals. A data set may look like this:

| id | parity | age | year | next.ivl | event | prev.ivl | ses |
|----|--------|-----|------|----------|-------|----------|-----|
| 1 | 0 | 24 | 1825 | 0.411 | 1 | NA | farmer |
| 1 | 1 | 25 | 1826 | 22.348 | 0 | 0.411 | farmer |
| 2 | 0 | 18 | 1821 | 0.304 | 1 | NA | unknown |
| 2 | 1 | 19 | 1821 | 1.837 | 1 | 0.304 | unknown |
| 2 | 2 | 21 | 1823 | 2.546 | 1 | 1.837 | unknown |
| 2 | 3 | 23 | 1826 | 2.541 | 1 | 2.546 | unknown |
| 2 | 4 | 26 | 1828 | 2.431 | 1 | 2.541 | unknown |
| 2 | 5 | 28 | 1831 | 2.472 | 1 | 2.431 | unknown |
| 2 | 6 | 31 | 1833 | 3.173 | 0 | 2.472 | unknown |

This is the first 9 rows, corresponding to the first two mothers in the data file. The variable *id* is *mother's id*, a label that uniquely identifies each individual.

A birth interval has a start point (in time) and an end point. These points are the time points of births, except for the first interval, where the start point is time of marriage, and the last interval, which is open to the right. However, the last interval is stopped at the time of marriage dissolution or when the mother becomes 50, whatever comes first. The variable *parity* is zero for the first interval, between date of marriage and date of first birth, one for the next interval, and so forth. The last (highest) number is thus equal to the total number of births for a woman during her first marriage (disregarding twin births, etc.).

Here is a description variable by variable of the data set.

- **id** The mother's unique id.
- **parity** Order of *previous* birth, see above for details. Starts at zero.
- **age** Mother's age at the event defining the start of the interval.
- **year** Calendar year for the birth defining the start of the interval.
- **next.ivl** measures the time in years from the birth at *parity* to the birth at *parity + 1*, or, for the woman's last interval, to the age of right censoring.
- **event** is an indicator for the interval ending with a birth. It is always equal to 1, except for the last interval, which always has event equal to zero.
- **prev.ivl** is the length of the interval preceding this one. For the first interval of a woman, it is always *NA* (Not Available).
- **ses** Socio-economic status (based on occupation data).

Just to make it clear: The first woman has id 1. She is represented by two records, meaning that she gave birth to one child. She waited 0.411 years from marriage to the first birth, and 22.348 years from the first birth to the second, *which never happened*. The second woman (2) is represented by seven records, implying that she gave birth to six children. And so on.

Of course, in an analysis of birth intervals we are interested in causal effects; why are some intervals short while others are long? The dependence of the history can be modeled by *lengths of previous intervals* (for the same mother), *parity, survival of earlier births*, and so on. Note that all relevant covariate information *must refer to the past*. More about that later.

The first interval of a woman is different from the others, since it starts with marriage. It therefore makes sense to analyze these intervals separately. The last interval of a woman is also special; it always ends with a right censoring, at the latest when the woman is 50 years of age. You should think of data for a woman generated sequentially in time, starting at the day of her marriage. Follow-up is made to the next birth, as long as she is alive, the marriage is still alive, and she is younger than 50 years of age. If there is no next birth, i.e., she reaches 50, or the marriage is dissolved (most often by death of one of the spouses), the interval is censored at the duration when she still was under observation. Censoring can also occur by emigration, and reaching the end of follow-up, in this case November 5, 1901. □

**Example 1.3** (The illness-death model)

Another useful setup is the *illness-death* model, see Figure 1.5.

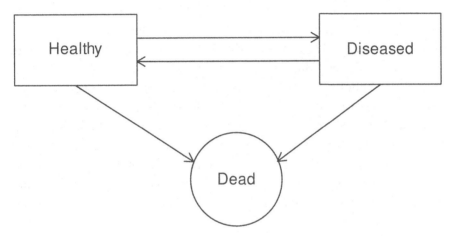

**FIGURE 1.5** The illness-death model

Individuals may move back and forth between the states *Healthy* and *Diseased*, and from each of these two states there is a pathway to *Dead*, which is an *absorbing state*, meaning that once in that state, you never leave it. □

## 1.6 More Data Sets

A few examples and data sets will be used repeatedly throughout the book, and we give a brief description of them here. They are all available in the R package **eha**, which is loaded into a running R session by the call

```
library(eha)
```

This loads the **eha** package. In the examples to follow, we assume
that this is already done. The main data source is the CEDAR,
Umeå University, Sweden. However, one data set is taken from the
home page of Statistics Sweden[3].

**Example 1.4** (Survival of males aged 20)

This data set is included in the R (R Core Team, 2021) package
**eha** (Broström, 2021). It contains information about 1023 males,
age twenty between January 1, 1800 and December 31, 1819, and
living in Skellefteå, a parish in the north-east of Sweden. The
total number of records in the data frame is 1211, that is, some
individuals are represented by more than one record in the data file.
The reason for that is that the *socio-economic status* (ses) is one
of the covariates in the file, and it changes over time. Each time
a change is recorded, a new record is created for that individual,
with the new value of SES. For instance, the third and fourth rows
in the data frame are

```
id  enter    exit event birthdate    ses
 3  0.000  13.463     0  1800.031  upper
 3 13.463  20.000     0  1800.031  lower
```

Note that the variable id is the same (3) for the two records,
meaning that both records are information about individual No. 3.
The variable enter is age (in years) that has elapsed since the 20th
birth day anniversary, and exit likewise. The information about
him is that he was born on 1800.031, or January 12, 1800, and
he is followed from his 21th birth date, or from January 12, 1820.
He is in an upper socio-economic status until he is 20 + 13.463 =
33.463 years of age, when he unfortunately is degraded to a lower
ses. He is then followed until 20 years have elapsed, or until his

---

[3]https://www.scb.se

fortieth birthday. The variable event tells us that he is alive we stop observing him. The value zero indicates that the follow-up ends with *right censoring*.

In an analysis of male mortality with this data set we could ask whether there is a socio-economic difference in mortality, and also if it changes over time. That would typically be done by *Cox regression* or by a parametric *proportional hazards model*. More about that follows in later chapters.

**Example 1.5** (Child mortality)

The data set child (in eha) contains follow-up data of children born in Skellefteå, northern Sweden, between January 1, 1850 and December 31, 1884. Each child is followed up to a maximum of fifteen years, and the follow-up is interrupted by death, loss of follow-up (usually out-migration), or the reach of age fifteen.

These data constitute a *cohort* and can be analyzed as such. It may be represented in a *Lexis diagram* as in Figure 1.6.

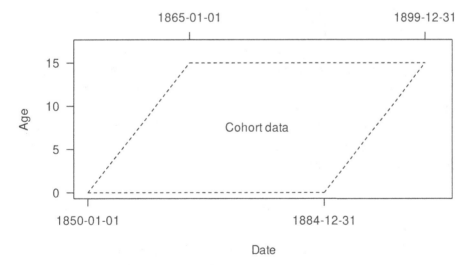

**FIGURE 1.6** Birth cohort 1850–1884, Skellefteå. A child mortality study.

The first few lines are

| id | m.id | sex | socBranch | birthdate | exit | event | illeg | m.age |
|---|---|---|---|---|---|---|---|---|
| 9 | 246606 | male | farming | 1853-05-23 | 15.000 | 0 | no | 35.009 |
| 150 | 377744 | male | farming | 1853-07-19 | 15.000 | 0 | no | 30.609 |
| 158 | 118277 | male | worker | 1861-11-17 | 15.000 | 0 | no | 29.320 |
| 178 | 715337 | male | farming | 1872-11-16 | 15.000 | 0 | no | 41.183 |
| 263 | 978617 | female | worker | 1855-07-19 | 0.559 | 1 | no | 42.138 |

Not that the variable enter really is redundant here: It is always equal to zero since all children are followed from birth in Skellefteå.

This data set will be discussed in more detail in a later chapter. It is part of a study of child mortality  □

**Example 1.6** (Infant mortality)

This data set is taken from Broström (1987) and concerns the interplay between infant and maternal mortality in 19th century Sweden (source: CEDAR, Umeå University, Sweden). More specifically, we are interested in estimating the effect of mother's death on the infant's survival chances. Because maternal mortality was rare (around one per 200 births), matching is used. This is performed as follows: for each child experiencing the death of its mother (before age one), two matched controls were selected. The criteria were: same age as the case at the event, same sex, birth year, parish, socio-economic status, marital status of mother. The triplets so created were followed until age one, and eventual deaths of the infants were recorded. The data collected in this way is part of the **eha** package under the name *infants*, and the first rows of the data frame are shown here:

| id | enter | exit | event | mother | age | sex | parish | civst | ses | year |
|---|---|---|---|---|---|---|---|---|---|---|
| 1 | 55 | 365 | 0 | dead | 26 | boy | Nedertornea | married | farmer | 1877 |
| 1 | 55 | 365 | 0 | alive | 26 | boy | Nedertornea | married | farmer | 1870 |
| 1 | 55 | 365 | 0 | alive | 26 | boy | Nedertornea | married | farmer | 1882 |

A short description of the variables follows.

- **id** denotes the id of the mother, 35 in all.

- **enter** is the age in days of the case, when its mother died.
- **exit** is the age in days when follow-up ends. It takes the value 365 (one year) for those who survived their first anniversary.
- **event** indicates whether a death (1) or a survival (0) was observed.
- **mother** has value *dead* for all cases and the value *alive* for the controls.
- **age** Age of mother at infant's birth.
- **sex** Sex of the infant.
- **parish** Birth parish.
- **civst** Civil status of mother, married or unmarried.
- **ses** Socio-economic status, often the father's, based on registrations of occupation.
- **year** Calendar year of the birth.

This data set is discussed and analyzed in Chapter 8. □

**Example 1.7** (Mortality in Sweden 1969–2020, tabular data)

These data sets (swepop and swedeaths) are taken from Statistics Sweden, freely available on the web site https://www.scb.se. The aggregated data sets contain yearly information about population size and number of deaths by sex and age for the years 1969–2020.

The first and last rows of swepop for the year 2020 are

```
age    sex year        pop
  0  women 2020  55505.5
  0    men 2020  58980.5
  1  women 2020  57158.0


age    sex year      pop
 99    men 2020    367.0
100  women 2020   1933.5
100    men 2020    394.5
```

and the first and last rows of swedeaths are

| age | sex | year | deaths |
|-----|-----|------|--------|
| 0 | women | 2020 | 112 |
| 0 | men | 2020 | 156 |
| 1 | women | 2020 | 8 |

| age | sex | year | deaths |
|-----|-----|------|--------|
| 99 | men | 2020 | 178 |
| 100 | women | 2020 | 1004 |
| 100 | men | 2020 | 243 |

The variables have the following meanings.

- **pop** Average population size 2007 in the *age* and for the *sex* given on the same row. The average is based on the population at the beginning and end of the year 2007.
- **deaths** The observed number of deaths in the *age* and for the *sex* given on the same row.
- **sex** Female or male.
- **age** Age in completed years.

See Chapter 7 for how to analyze this data set.  □

# 2

## Single Sample Data

The basic model descriptions of survival data are introduced. Basically, the distribution of survival data may be either *continuous* or *discrete*, but the nonparametric estimators of the distributions of survival time are discrete in any case. Therefore, the discrete models are necessary to know in order to understand nonparametric estimation.

In the present chapter, only nonparametric estimation is discussed. This means that no assumptions whatsoever are made about the true underlying distribution. Parametric models are presented in Chapter 8.

## 2.1 Continuous Time Model Descriptions

Traditional statistical model descriptions are based on the *density* and the *cumulative distribution* functions. These functions are not so suitable for use with censored and/or truncated data. The *survival* and *hazard functions* are better suited, as will be demonstrated here. It should however be acknowledged that all these functions are simple functions of each other, so the information they convey is in principle the same, it is only convenience that determines which one(s) to choose and use in a particular application.

DOI: 10.1201/9780429503764-2

### 2.1.1   The survival function

We start with a motivating example (2.1), the life table, and its
connection to death risks and survival probabilities in a population,
see Table 2.1.

**Example 2.1** (Swedish life tables)

Statistics Sweden (SCB) is a government agency that produces
statistics and have a coordinating role for the official statistics of
Sweden. From their home page[1] it is possible to download vital
population statistics, like population size and number of deaths by
age, gender, and year. From such data it is possible to construct
*life tables*. Two raw tables are needed, one with population size
by age, sex, and year, and one with the number of deaths by the
same categories. Tables are available for age intervals of lengths 1,
5, and 10 years. We always prefer data with as much information
as possible, so we choose the tables with age interval lengths one.
It is always possible to *aggregate* later, if that is preferred.

There are two data sets in the **R** package **eha** that contain population size information and information about number of deaths,
swepop and swedeaths, respectively. Both cover the years 1969–2020.
We use both to construct a data frame females containing population size and number of deaths for females 2020. Below, the first
five and last five rows are printed. Note that the last row, with age
= 100 in fact includes all ages from 100 and above. It is common
that tables like this one ends with an open interval.

```
library(eha)
females <- swepop[swepop$sex == "women" & swepop$year == 2020,
                  c("age", "pop")]
females$deaths <- swedeaths[swedeaths$sex == "women" &
                            swedeaths$year == 2020, "deaths"]
print(females[c(1:5, 97:101), ], row.names = FALSE)
```

---
[1] https://www.scb.se

| age | pop | deaths |
|---:|---:|---:|
| 0 | 55505.5 | 112 |
| 1 | 57158.0 | 8 |
| 2 | 58254.5 | 5 |
| 3 | 59680.0 | 3 |
| 4 | 60075.5 | 9 |
| 96 | 3940.5 | 1256 |
| 97 | 2768.0 | 883 |
| 98 | 2033.5 | 764 |
| 99 | 1445.0 | 581 |
| 100 | 1933.5 | 1004 |

A full view of the population size by age is given in Figure 2.1.

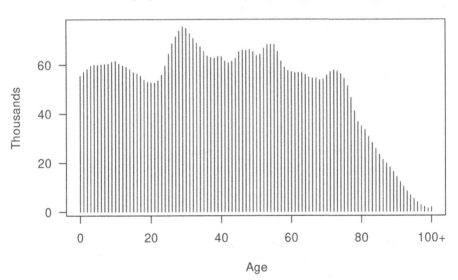

**FIGURE 2.1** Female population by age, Sweden 2020.

And the numbers of deaths by age are shown in Figure 2.2.

Our primary interest lies in the *age specific mortality*, which simply is given by dividing the number of deaths by the population size, age by age. We do that, and add one column, derived from the original three, in the following way.

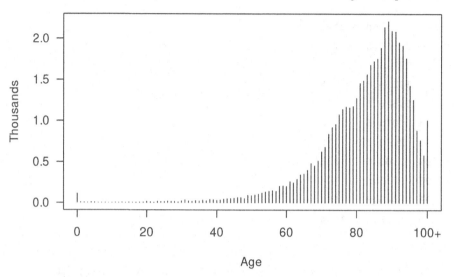

**FIGURE 2.2** Female deaths by age, Sweden 2020.

```
females$risk <- females$deaths / females$pop
alive <- numeric(101)
alive[1] <- 100000
for (i in 2:101){
    alive[i] <- alive[i - 1] * (1 - females$risk[i-1])
}
females$alive <- alive
```

The result is shown in Table 2.1.

It is constructed in the following way. The first column, headed by *age*, contains the ages in completed years. The second column, headed *pop* contains the numbers of women in Sweden in the corresponding ages. It is equal to the mean population size over the year. Since population numbers from the SCB are taken at the turn of the years, the numbers in column two are averages of two columns in the SCB file. The third column contains the total numbers of deaths in each age category during the year. The fourth and fifth columns are derived from the previous ones. The *risk* is

**TABLE 2.1** Life table for Swedish women 2020.

| age | pop | deaths | risk | alive |
|---|---|---|---|---|
| 0 | 55506 | 112 | 0.00202 | 100000 |
| 1 | 57158 | 8 | 0.00014 | 99798 |
| 2 | 58254 | 5 | 0.00009 | 99784 |
| 3 | 59680 | 3 | 0.00005 | 99776 |
| 4 | 60076 | 9 | 0.00015 | 99771 |
| 96 | 3940 | 1256 | 0.31874 | 7981 |
| 97 | 2768 | 883 | 0.31900 | 5437 |
| 98 | 2034 | 764 | 0.37571 | 3703 |
| 99 | 1445 | 581 | 0.40208 | 2312 |
| 100+ | 1934 | 1004 | 0.51927 | 1382 |

the ratio between the number of deaths and population size for each age.

Column five, *alive*, contains the actual construction of the life table. It depicts the actual size of a hypothetical birth cohort, *year by year*, exposed to the death risks of column four. It starts with 100000 (an arbitrary, but commonly chosen, number) newborn females. Under the first year of life the death risk is 0.00202, meaning that we expect 0.00202 x 100000 = 202 girls to die in the first year. In other words, 100000−202 = 99798 will remain alive at the first anniversary (at exact age one year). That is the number in the second row in the fifth column. The calculations then continues in the same way through the last two columns. Note, though, that the last interval (100+) is an open interval, i.e., it has no fixed upper limit.

Note the difference in interpretation between the numbers in the second column ("pop") and those in the last ("alive"); the former gives the age distribution in a population a given year (referred to as *period data*), while the latter constitutes synthetic *cohort data*, i.e., a group of newborn are followed from birth to death, and the actual size of the (still alive) cohort is recorded age by age. That is, "individuals" in the synthetic cohort live their whole lives under the mortality conditions of the year 2020. Of course no real individual can experience this scenario, hence the name "synthetic cohort". Nevertheless, it is a useful measure for illustrating the state of affairs regarding mortality a given year.

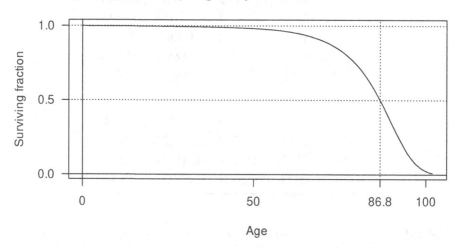

**FIGURE 2.3** Life table and survival function, females, Sweden 2020.

It makes sense to plot the life table, see Figure 2.3. Exactly how this is done with individual data will be shown later in the chapter, but here we simply plot alive (divided by 100000) against age. *Median life length* is 86.8, see construction in the figure. By dividing all numbers by the original cohort size (in our case 100000), they may be interpreted as *probabilities*. □

The *survival function* $S(t)$, $t > 0$, is defined as the probability of surviving past $t$, $t > 0$, or with a formula,

$$S(t) = P(T \geq t), \quad t > 0, \tag{2.1}$$

where $T$ is the (random) life length under study. The symbol $P(T \geq t)$ reads *"the probability that $T$ is equal to or larger than $t$"*; here $t$ is a fixed number while $T$ denotes a "random quantity". In statistical language, $T$ is called a *random variable*. In our example, $T$ is the (future) life length of a randomly chosen newborn female, of course unknown at birth.

We will, for the time being, assume that $S$ is a "nice" function, smooth and *differentiable*. That will ensure that the following definitions are unambiguous.

### 2.1.2 The density function

The *density function* is defined as minus the derivative of the survival function, or

$$f(t) = -\frac{d}{dt}S(t), \quad t > 0 \tag{2.2}$$

The intuitive interpretation of this definition is that, for small enough $s$, the following approximation is good:

$$P(t_0 \leq T < t_0 + s) = S(t_0) - S(t_0 + s) \approx s f(t_0).$$

This is illustrated in Figure 2.4, with $t_0 = 10$ and $s = 2$.

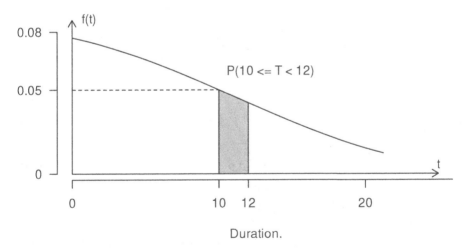

Duration.

**FIGURE 2.4** Interpretation of the density function.

For a short enough interval $(t_0, t_0 + s]$, the probability of an observation falling in that interval is well approximated by the area of a rectangle with sides of lengths $s$ and $f(t_0)$, or $s f(t_0)$. The formal mathematical definition as a limit is given in equation (2.3).

$$f(t) = \lim_{s \to 0} \frac{P(t \le T < t + s)}{s}, \quad t > 0. \tag{2.3}$$

For the Swedish women 2020, the estimated density function is
shown in Figure 2.5.

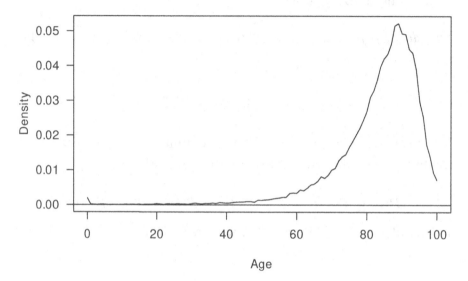

**FIGURE 2.5** Density function for the life length of Swedish
women 2020.

### 2.1.3 The hazard function

The *hazard function* is central to the understanding of survival
analysis, so you are recommended to get at least an intuitive
understanding of it. One way of thinking of it is as an "instant
probability" (per unit time); at a given age $t$, it measures the risk
of dying in a short interval $(t, t + s)$ immediately after $t$, for an
individual *who still is alive at t.*

$$h(t) = \lim_{s \to 0} \frac{P(t \le T < t + s \mid T \ge t)}{s}$$

$$= \lim_{s \to 0} \frac{S(t) - S(t+s)}{sS(t)}$$

$$= \frac{f(t)}{S(t)}$$

Note the difference between the density and the hazard functions. The former is (the limit of) an *unconditional* probability, while the latter is (the limit of) a *conditional* probability per time unit.

For the Swedish women 2020, the estimated hazard function (age-specific mortality) is shown in Figure 2.6.

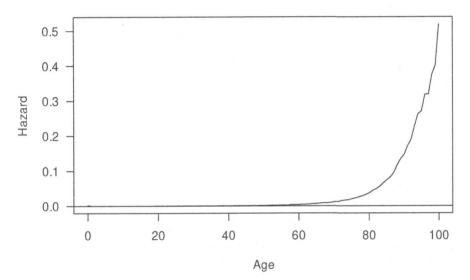

**FIGURE 2.6** Hazard function for the life length of Swedish women 2020.

### 2.1.4    The cumulative hazard function

The *cumulative hazard function* is defined as the integral of the hazard function,

$$H(t) = \int_0^t h(s)ds, \quad t \ge 0.$$

That is, an intuitive interpretation is that the cumulative hazard function successively accumulates the instant risks.

The cumulative hazard function is important because it is fairly easy to estimate nonparametrically (i.e., without any restrictions or assumptions), in contrast to the hazard and density functions.

For the Swedish women 2020, the estimated cumulative hazard function is shown in Figure 2.7.

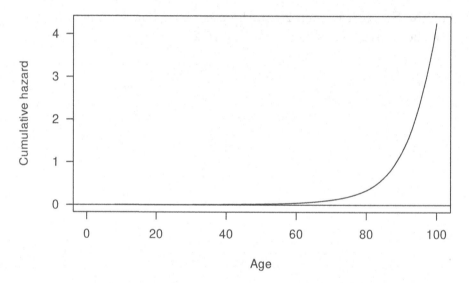

**FIGURE 2.7** Cumulative hazard function for Swedish women 2020.

**Example 2.2** (The exponential distribution)

Perhaps the simplest continuous life length distribution is the *exponential distribution*. It is simple because its hazard function is constant:

$$h(t) = \lambda, \quad \lambda > 0, \ t \ge 0.$$

From this it is easy to calculate the other functions that characterize the exponential distribution. The cumulative hazards function is

$$H(t) = \lambda t, \quad \lambda > 0, \ t \geq 0,$$

the survival function is

$$S(t) = e^{-\lambda t}, \quad \lambda > 0, \ t \geq 0,$$

and the density function is

$$f(t) = \lambda e^{-\lambda t}, \quad \lambda > 0, \ t \geq 0.$$

The property of constant hazard implies *no aging*. This is not a realistic property for human mortality, but, as we will see, a useful benchmark, and a useful tool for modeling mortality over short time intervals (*piecewise constant hazard*). The exponential distribution is described in detail in Appendix B. □

## 2.2   Discrete Time Models

So far we have assumed, implicitly or explicitly, that time is continuous. We will now introduce discrete time survival models, and the reason is two-fold: (i) Even if data are generated from truly continuous time models, nonparametric estimation of these models will, as will be shown later, give rise to estimators corresponding to a discrete time model. This is an inherent property of nonparametric maximum likelihood estimation. Thus, in order to study the properties of these estimators, we need some knowledge of discrete time models. (ii) Data are discrete, usually through grouping. For instance, life lengths may be measured in full years, introducing *tied data*.

It is important to realize that in practice all data are discrete. For instance, it is impossible to measure time with infinite precision. Therefore, all data are more or less rounded. If data are so much rounded that the result is heavily tied data, true discrete-data models are called for.

Discrete time models will now be introduced. Let $R$ be a discrete random variable with

- support $(r_1, r_2, \ldots, r_k)$ (positive real numbers, usually $1, 2, \ldots$ or $0, 1, 2, \ldots$),
- probability mass function

$$p_i = P(R = r_i), \quad i = 1, \ldots, k,$$

with $p_i > 0, \quad i = 1, \ldots, k$ and $\sum_{i=1}^{k} p_i = 1$.

Then

$$F(t) = \sum_{i:r_i \leq t} p_i, \quad -\infty < t < \infty,$$

is the cumulative distribution function, and

$$S(t) = \sum_{i:r_i \geq t} p_i, \quad -\infty < t < \infty,$$

is the survival function.

The discrete time hazard function is defined as

$$h_i = P(R = r_i \mid R \geq r_i) = \frac{p_i}{\sum_{j=i}^{k} p_j}, \quad i = 1, \ldots, k. \qquad (2.4)$$

Note that here, the hazard at any given time point is a *conditional probability*, so it must always be bounded to lie between zero and one. In the continuous case, on the other hand, the hazard function may take any positive value. Further note that if, like here, the support is finite, the last "hazard atom" is always equal to one (having lived to the "last station", one is bound to die).

The system (2.4) of equations has a unique solution, easily found by recursion:

$$p_i = h_i \prod_{j=1}^{i-1}(1 - h_j), \quad i = 1, \dots, k. \tag{2.5}$$

From this we get the discrete time survival function at each support point as

$$S(r_i) = \sum_{j=i}^{k} p_j = \prod_{j=1}^{i-1}(1 - h_j), \quad i = 1, \dots, k,$$

and the general definition

$$S(t) = \prod_{j:r_j < t} (1 - h_j), \quad t \geq 0 \tag{2.6}$$

It is easily seen that $S$ is decreasing, $S(0) = 1$, and $S(\infty) = 0$, as it should be.

**Example 2.3** (The geometric distribution)

The *geometric* distribution has support on $\{1, 2, \dots\}$ (another version also includes zero in the support, this is the case for the one in **R**), and the hazard function $h$ is constant:

$$h_i = h, \quad 0 < h < 1, \ i = 1, 2, \dots.$$

Thus, the geometric distribution is the discrete analogue to the exponential distribution in that it implies no aging. The probability mass function is, from (2.5),

$$p_i = h(1 - h)^{i-1}, i = 1, 2, \dots,$$

and the survival function becomes, from (2.6),

$$S(t) = (1 - h)^{[t]}, \quad t > 0,$$

where $[t]$ denotes the largest integer smaller than or equal to $t$ (rounding toward zero).

## 2.3  Nonparametric Estimators

As an introductory example, look at an extremely simple data set: 4, 2\*, 6, 1, 3\* (starred observations are right censored; in the figure, deaths are marked with +, censored observations with a small circle), see Figure 2.8.

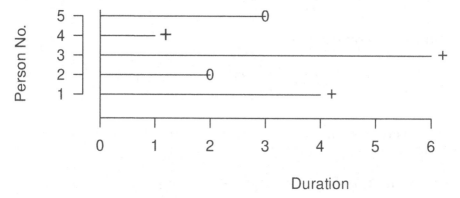

**FIGURE 2.8** A simple survival data set.

How should the survival function be estimated? The answer is that we take it in steps. First, the hazard "atoms" are estimated. It is done nonparametrically, and the result as such is not very useful. Its potential lies in that it is used as the building block in constructing estimates of the cumulative hazards and survival functions.

### 2.3.1  The hazard atoms

In Figure 2.9, the observed event times in Figure 2.8 are marked by the vertical dashed lines at durations 1, 4, and 6, respectively. In the estimation of the hazard atoms, the concept of *risk set* is of vital importance.

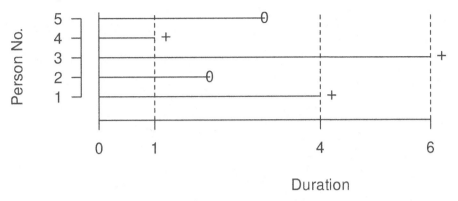

Duration

**FIGURE 2.9** Preliminaries for estimating the hazard function.

The risk set $R(t)$ at duration $t$, $t > 0$ is defined mathematically as

$$R(t) = \{\text{all individuals under observation at } t-\}, \qquad (2.7)$$

or in words, the risk set at time $t$ consists of all individuals present and under observation just prior to $t$. The reason we do not say "present at time $t$" is that it is vital to include those individuals who have an event or are right censored at the exact time $t$. In our example, the risk sets $R(1)$, $R(4)$, and $R(6)$ are the interesting ones. They are:

$$R(1) = \{1, 2, 3, 4, 5\}$$
$$R(4) = \{1, 3\}$$
$$R(6) = \{3\}$$

The estimation of the hazard atoms is simple. First, we assume that the probability of an event at times where no event is observed, is zero. Then, at times where events do occur, we count the number of events and divides that number by the size of the corresponding risk set. The result is shown in (2.8).

$$\hat{h}(1) = \frac{1}{5} = 0.2$$

$$\hat{h}(4) = \frac{1}{2} = 0.5 \qquad (2.8)$$

$$\hat{h}(6) = \frac{1}{1} = 1$$

See also Figure 2.10.

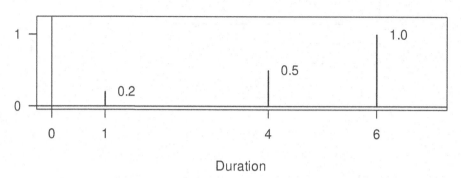

Duration

**FIGURE 2.10** Nonparametric estimation of the hazard function.

As is evident from Figure 2.10, the estimated hazard atoms will be too irregular to be of practical use; they need *smoothing*. The simplest way of smoothing them is to calculate the cumulative sums, which leads to the *Nelson-Aalen* estimator (Nelson, 1972; Aalen, 1978) of the cumulative hazards function, see the next section 2.3.2. There are more direct smoothing techniques to get reasonable estimators of the hazard function itself, e.g., kernel estimators , but they will not be discussed here. See e.g. Silverman (1986) for a general introduction to kernel smoothing.

### 2.3.2   The Nelson-Aalen estimator

From the theoretical relation we immediately get

$$\hat{H}(t) = \sum_{s \le t} \hat{h}(s), \quad t \ge 0,$$

which is the *Nelson-Aalen* estimator (Nelson, 1972; Aalen, 1978), see Figure 2.11. The sizes of the jumps are equal to the heights of the "spikes" in Figure 2.10.

### 2.3.3   The Kaplan-Meier estimator

From the theoretical relation (2.6) we get

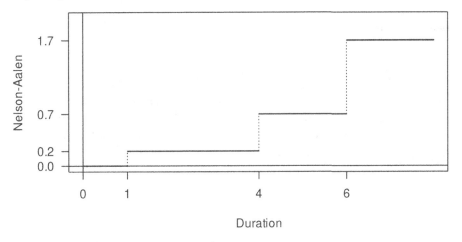

**FIGURE 2.11** The Nelson-Aalen estimator.

$$\hat{S}(t) = \prod_{s<t}(1 - \hat{h}(s)), \quad t \geq 0, \tag{2.9}$$

see also Figure 2.12. Equation (2.9) may be given a heuristic interpretation: In order to survive time $t$, one must survive *all* "spikes" (or shocks) that come before time $t$. The multiplication principle for conditional probabilities then gives equation (2.9).

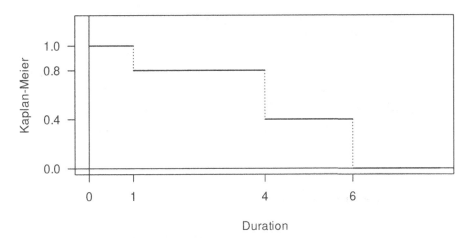

**FIGURE 2.12** The Kaplan-Meier estimator.

## 2.4 Doing It in R

We show how to do it in **R** by using the male mortality data set
mort, which is part of the package eha. By loading eha we have
access to mort. The **R** function head is convenient for looking at a
few (six by default) rows in a data frame.

```
library(eha)
head(mort)
```

|   | id | enter  | exit   | event | birthdate | ses   |
|---|----|--------|--------|-------|-----------|-------|
| 1 | 1  | 0.000  | 20.000 | 0     | 1800.010  | upper |
| 2 | 2  | 3.478  | 17.562 | 1     | 1800.015  | lower |
| 3 | 3  | 0.000  | 13.463 | 0     | 1800.031  | upper |
| 4 | 3  | 13.463 | 20.000 | 0     | 1800.031  | lower |
| 5 | 4  | 0.000  | 20.000 | 0     | 1800.064  | lower |
| 6 | 5  | 0.000  | 0.089  | 0     | 1800.084  | lower |

This is the normal form of storing left truncated and right censored
data. Here, males are followed from their twentieth birthdate until
death or their fortieth birthdate, whichever comes first. An indicator
for *death* is introduced, called event. The value 1 indicates that the
corresponding life time is fully observed, while the value 0 indicates
a right censored life time. Another common coding scheme is TRUE
and FALSE, respectively.

### 2.4.1 Nonparametric estimation

In the **R** package eha, the plot function can be used to plot both
Nelson-Aalen (cumulative hazards) and Kaplan-Meier (survival)
curves. Here is the code:

```
par(mfrow = c(1, 2))# Two panels, "one row, two columns".
with(mort, plot(Surv(enter, exit, event), fun = "cumhaz",
     main = "Cumulativa hazards function",
    xlab = "Duration"))
with(mort, plot(Surv(enter, exit, event),
               main = "Survival function",
              xlab = "Duration"))
```

and the result is seen in Figure 2.13.

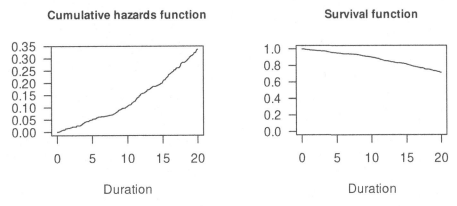

**FIGURE 2.13** Nelson-Aalen plot (left panel) and Kaplan-Meier plot (right panel), male mortality data.

Note the use of the function `with`; it tells the `plot` function that it should get its data (`enter`, `exit`, `event`) from the data frame `mort`. The function `Surv`, imported from the package `survival`, creates a "survival object", which is used in many places. It is for instance *the response* in all functions that perform regression analysis on survival data. Note that the "Duration" in Figure 2.13 is duration (in years) since the day each man became twenty years of age. They are followed until death or age forty, whichever comes first. The right hand panel shows that approximately 25 percent of the men alive at age twenty died before they became forty. Data come from a 19th century Swedish sawmill area.

### 2.4.2  Parametric estimation

It is also possible to fit a parametric model to data with the aid of the function `phreg` (it also works with the function `aftreg`). Just fit a "parametric proportional hazards model with no covariates".

```
par(mfrow = c(1, 2), las = 1)
fit.w <- phreg(Surv(enter, exit, event) ~ 1, data = mort)
plot(fit.w, fn = "cum", main = "")
plot(fit.w, fn = "sur", main = "")
```

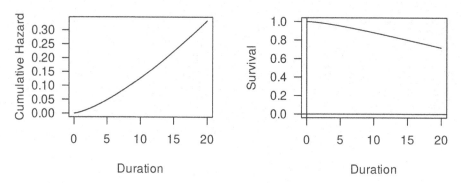

**FIGURE 2.14** Male mortality, Weibull fit.

Note that the default distribution in `phreg` is the *Weibull* distribution, which means that if no distribution is specified (through the argument `dist`), then the Weibull distribution is assumed.

# 3

## Proportional Hazards and Cox Regression

*Proportional hazards* is a property of survival models that is fundamental for the development of non-parametric regression models.
After defining this property, we move slowly from the two-sample
case, via the $k$-sample case, to the general regression model. On
this route, we start with the *log-rank test* and end up with *Cox
regression* (Cox, 1972).

### 3.1 Proportional Hazards

The property of *proportional hazards* is fundamental in Cox regression. It is in fact the essence of Cox's simple, yet ingenious idea.
The definition is as follows:

**Definition 3.1** (Proportional hazards) If $h_1(t)$ and $h_0(t)$ are hazard functions from two separate distributions, we say that they are
*proportional* if

$$h_1(t) = \psi h_0(t), \quad \text{for all } t \geq 0, \tag{3.1}$$

for some positive constant $\psi$ and *all $t \geq 0$*. Further, if (3.1) holds,
then the same property holds for the corresponding *cumulative
hazard functions* $H_1(t)$ and $H_0(t)$.

$$H_1(t) = \psi H_0(t), \quad \text{for all } t \geq 0, \tag{3.2}$$

with the same proportionality constant $\psi$ as in (3.1). $\square$

Strictly speaking, the second part of this definition follows easily

DOI: 10.1201/9780429503764-3

from the first (and vice versa), so more correct would be to state one part as a definition and the other as a corollary. The important part of this definition is *"for all $t \geq 0$"*, and that the constant $\psi$ *does not depend on t.*

Think of the hazard functions as age-specific mortality for two groups, e.g., women and men. It is "well known" that women have lower mortality than men in all ages. It would therefore be reasonable to assume proportional hazards in that case. It would mean that the female *relative* advantage is equally large in all ages. See Example 3.1 for an investigation of this statement.

**Example 3.1** (Male and female mortality, Sweden 2001−2020)

This is a continuation of Example 2.1, and here we involve the male mortality alongside with the female, see Figure 3.1. We also expand the calendar time period to the years 2001–2020. The reason for this is that random fluctuations in low ages, caused by the extremely low mortality rates, blur the picture.

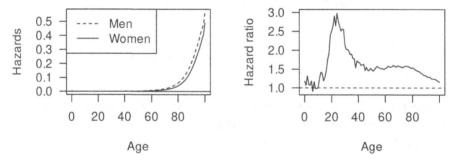

**FIGURE 3.1** Female and male mortality by age, Sweden 2001−2020. The right hand panel shows the ratio between the male and female mortality by age.

As you can see, the "common knowledge" of proportional hazards is disproved by the right hand panel in Figure 3.1. On the other hand, the left hand panel gives a rather different impression, and this illustrates the need to choose good approaches to graphical presentation, depending on what you want to show.

It must be emphasized that the proportional hazards assumption is an assumption that always must be carefully checked. In many situations, it would not be reasonable to assume proportional hazards. If in doubt, check data by plotting the Nelson-Aalen estimates for each group in the same plot. The left hand panel of Figure 3.1 would suit this purpose better if drawn on a log scale, see Figure 3.2.

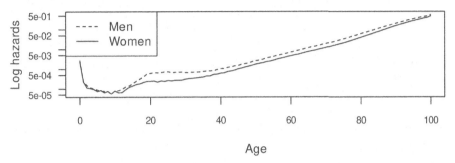

**FIGURE 3.2** Female and male mortality by age, Sweden 2001–2020. Log scale.

The advantage of the log scale is twofold: (i) Small numbers are magnified so you can see them, and (ii) on the log scale, proportional hazards implies a constant vertical distance between the curves, which is easier to see for the human eye.

☐ For an example of a perfect fit to the proportional hazards model, see Figure 3.3 (two *Weibull* hazard functions with the proportionality constant $\psi = 2$).

In the right hand panel of Figure 3.3, note that both dimensions are on a log scale. This type of plot, constructed from empirical data, is called a *Weibull plot* in reliability applications: If the lines are straight lines, then data are well fitted by a Weibull distribution. Additionally, if the the slope of the line is 1 (45 degrees), then an exponential model fits well.

To summarize Figure 3.3: (i) The hazard functions are proportional because on the log-log scale, the vertical distance is constant, (ii) Both hazard functions represent a Weibull distribution, because both lines are straight lines, and (iii) neither represents an exponential distribution, because the slopes are *not* one. This latter

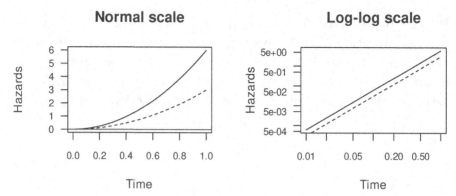

**FIGURE 3.3** Two hazard functions that are proportional. The proportionality constant is 2.

fact may be difficult to see because of the different scales on the axes (the *aspect ratio* is not one).

Figure 3.4 shows the relationships between the cumulative hazards functions, the density functions, and the survival functions when the hazard functions are proportional. Note that the cumulative hazards functions are proportional by implication, with the same proportionality constant ($\psi = 2$ in this case). On the other hand, for the density and survival functions, proportionality does not hold; it is in fact theoretically impossible except in the trivial case that the proportionality constant is unity.

## 3.2   The Log-Rank Test

The *log-rank test* is a $k$-sample test of equality of survival functions. It is a powerful test against proportional hazards alternatives, but may be very weak otherwise. We first look at the two-sample case, that is, $k = 2$.

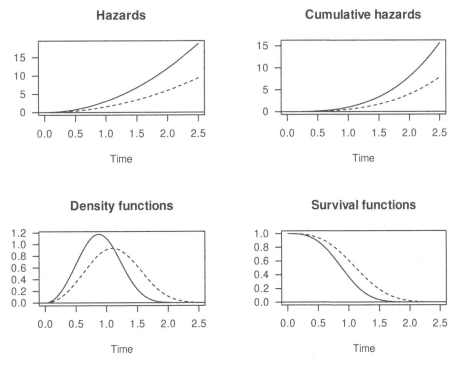

**FIGURE 3.4** The effect of proportional hazards on the density and survival functions.

### 3.2.1 Two samples

Suppose that we have the small data set illustrated in Figure 3.5. There are two samples, the letters (A, B, C, D, E) and the numbers (1, 2, 3, 4, 5).

The data in Figure 3.5 can be presented i tabular form, see Table 3.1.

We are interested in investigating whether letters and numbers have the same survival chances or not. Therefore, the hypothesis

$H_0$ : No difference in survival between numbers and letters

is formulated. In order to test $H_0$, we make five tables, one for each observed *event time*, see Table 3.2, where the the first table, relating to failure time $t_{(1)} = 1$, is shown.

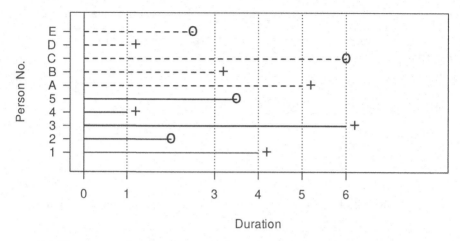

**FIGURE 3.5** Two-sample data, the `letters` (dashed) and the `numbers` (solid). Circles denote censored observations, plusses events.

**TABLE 3.1** Example data for the log-rank test.

| group | time | event |
|---|---|---|
| numbers | 4.0 | TRUE |
| numbers | 2.0 | FALSE |
| numbers | 6.0 | TRUE |
| numbers | 1.0 | TRUE |
| numbers | 3.5 | FALSE |
| letters | 5.0 | TRUE |
| letters | 3.0 | TRUE |
| letters | 6.0 | FALSE |
| letters | 1.0 | TRUE |
| letters | 2.5 | FALSE |

**TABLE 3.2** The 2x2 table for the first event time.

|  | Deaths | Survivals | Total |
|---|---|---|---|
| numbers | 1 | 4 | 5 |
| letters | 1 | 4 | 5 |
| Total | 2 | 8 | 10 |

**TABLE 3.3** Observed and expected number of deaths for numbers at event times.

|        | Observed | Expected | Difference | Variance |
|--------|----------|----------|------------|----------|
| $t(1)$ | 1        | 1.0      | 0.0        | 0.44     |
| $t(2)$ | 0        | 0.5      | −0.5       | 0.25     |
| $t(3)$ | 1        | 0.5      | 0.5        | 0.25     |
| $t(4)$ | 0        | 0.3      | −0.3       | 0.22     |
| $t(5)$ | 1        | 0.5      | 0.5        | 0.25     |
| Sum    | 3        | 2.8      | 0.2        | 1.41     |

Let us look at the table at failure time $t_{(1)} = 1$, Table 3.2, from the viewpoint of the numbers.

- The observed number of deaths among numbers: 1.
- The expected number of deaths among numbers: $2 \times 5/10 = 1$.

The *expected* number is calculated under $H_0$, i.e., as if there is no difference between letters and numbers regarding mortality. It is further assumed that the two margins (Total) are given (fixed) and that the number of deaths follows a hypergeometric distribution, which gives the variance.

Then, given two deaths in total and five out of ten observations are from the group numbers, the expected number of deaths is calculated as above.

This procedure is repeated for each of the five tables, and the results are summarized in Table 3.3.

Finally, the observed test statistic $T$ is calculated as

$$T = \frac{0.2^2}{1.41} \approx 0.028$$

Under the null hypothesis, this is an observed value from a $\chi^2(1)$ distribution, and $H_0$ should be rejected for *large* values of $T$. Using a *level of significance* of 5%, the cutting point for the value of $T$

**TABLE 3.4** Old age mortality.

| id | enter | exit | event | sex | civ |
|----|-------|------|-------|-----|-----|
| 709 | 66.498 | 67.988 | 0 | male | married |
| 709 | 67.988 | 72.820 | 0 | male | married |
| 709 | 72.820 | 75.542 | 1 | male | widow |
| 710 | 66.446 | 76.568 | 1 | female | married |
| 711 | 66.446 | 67.936 | 0 | female | married |

is 3.84, far from our observed value of 0.028. The conclusion is therefore that there is no (statistically significant) difference in survival chances between letters and numbers. Note, however, that this result depends on asymptotic (large sample) properties, and in this toy example, these properties are not valid.

For more detail about the underlying theory, see Appendix A.

Let us now look at a real data example, the old age mortality data set oldmort in eha. See Table 3.4 for a sample of five records with selected columns.

We are interested in comparing male and female mortality in the ages 60–85 with a log-rank test, and for that purpose we use the logrank function:

```
library(eha)
fit <- logrank(Surv(enter, exit, event), group = sex, data = om)
fit
```

```
        The log-rank test

Call:
logrank(Y = Surv(enter, exit, event), group = sex, data = om)

 X-squared =  19.1971 , df =  1 , p-value =  1.1789e-05
```

The result of the log-rank test is given by the $p$-value 0.0012 percent, a very small number. Thus, there is a very significant difference in mortality between men and women. But *how large* is the difference? The log-rank test has no answer to this question.

Remember that this result depends on the proportional hazards assumption. We can graphically check it, see Figure 3.6.

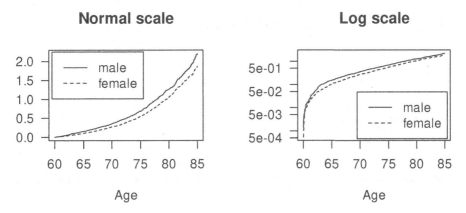

**FIGURE 3.6** Cumulative hazards for women and men in ages 60–85, 19th century Sweden.

The proportionality assumption appears to be a good description looking at the left hand panel, but it looks more doubtful on the log scale. It is an often observed phenomenon that differences in mortality between human groups, not only women and men, tend to vanish in the extremely high ages. The validity of our test result, however, is not affected by this deviation from proportionality, since the result was *rejection*. We can be confident in the fact that there is a significant difference, since violation of the proportionality assumption *in itself* implies a difference.

### 3.2.2 Several samples

The result for the two-sample case is easily extended to the $k$-sample case. Instead of one $2 \times 2$ table per observed event time we get one $k \times 2$ table per observed event time and we have to calculate expected and observed numbers of events for $(k - 1)$ groups at each failure time. The resulting test statistic will have $(k - 1)$ degrees of freedom and still be approximately $\chi^2$ distributed. This is illustrated with the same data set, `oldmort`, as above, but with the covariate `civ`, which is a `factor` with three levels (`unmarried`, `married`, `widow`), instead of `sex`. Furthermore, the investigation is limited to *male mortality*.

```
fit <- logrank(Surv(enter, exit, event), group = civ,
                    data = om[om$sex == "male", ])
fit
```

```
     The log-rank test

Call:
logrank(Y = Surv(enter, exit, event), group = civ, data = om[om$sex ==
    "male", ])

 X-squared =  19.8622 , df =  2 , p-value =  4.86372e-05
```

The degrees of freedom for the *score test* is now 2, equal to the number of levels in `civ` minus one. Figure 3.7 shows the cumulative hazards for each group. There is obviously nothing much that indicates non-proportionality in this case either. Furthermore, the ordering is what one would expect.

We do not go deeper into this matter here, mainly because the log-rank test is a special case of *Cox regression*.

## 3.3 Cox Regression Models

Starting with the definition of proportional hazards in Section 3.1, the concept of Cox regression is introduced in steps.

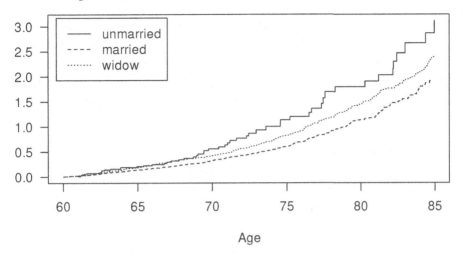

**FIGURE 3.7** Old age male mortality by civil status, cumulative hazards.

### 3.3.1 Two Groups

The definition of proportionality, given in equation (3.1), can equivalently be written as

$$h_x(t) = \psi^x h_0(t), \quad t > 0, \; x = 0, 1, \; \psi > 0. \tag{3.3}$$

It is easy to see that this "trick" is equivalent to (3.1): When $x = 0$, it simply says that $h_0(t) = h_0(t)$, and when $x = 1$ it says that $h_1(t) = \psi h_0(t)$. Since $\psi > 0$, we can calculate $\beta = \log(\psi)$, and rewrite (3.3) as

$$h_x(t) = e^{\beta x} h_0(t), \quad t > 0; \; x = 0, 1; \; -\infty < \beta < \infty,$$

or with a slight change in notation,

$$h(t; x) = e^{\beta x} h_0(t), \quad t > 0; \; x = 0, 1; \; -\infty < \beta < \infty. \tag{3.4}$$

The sole idea by this rewriting is to pave the way for the introduction of *Cox's regression model* (Cox, 1972), which in its elementary form is a proportional hazards model. In fact, we can already interpret equation (3.4) as a Cox regression model with an explanatory variable $x$ with corresponding regression coefficient $\beta$ (to be estimated from data). The covariate $x$ is still only a dichotomous variate, but we will now show how it is possible to generalize this to a situation with explanatory variables of any form. The first step is to go from the two-sample situation to a $k$-sample one.

### 3.3.2   Many groups

Can we generalize the results of Section 3.3.1 to $(k+1)$ groups, $k \geq 2$? Yes, by expanding that procedure as follows:

$$
\begin{aligned}
h_0(t) &\sim \text{group } 0 \\
h_1(t) &\sim \text{group } 1 \\
&\cdots \qquad \cdots \\
h_k(t) &\sim \text{group } k
\end{aligned}
$$

The underlying model is: $h_j(t) = \psi_j h_0(t), \quad t \geq 0, \ j = 1, 2, \dots, k$. That is, with $(k+1)$ groups, we need $k$ proportionality constants $\psi_1, \dots, \psi_k$ in order to define proportional hazards. Note also that in this formulation (there are others), one group is "marked" as a *reference group*, that is, to this group there is no proportionality constant attached. All relations are relative to the reference group. Note also that it doesn't matter which group is chosen as the reference. This choice does not change the model itself, only its representation. With $(k+1)$ groups, we need $k$ indicators. Let

$$
\mathbf{x} = (x_1, x_2, \dots, x_k).
$$

Then

$$
\begin{aligned}
\mathbf{x} &= (0, 0, \dots, 0) &&\Rightarrow \text{group } 0 \\
\mathbf{x} &= (1, 0, \dots, 0) &&\Rightarrow \text{group } 1 \\
\mathbf{x} &= (0, 1, \dots, 0) &&\Rightarrow \text{group } 2 \\
&\qquad \cdots &&\qquad \cdots \\
\mathbf{x} &= (0, 0, \dots, 1) &&\Rightarrow \text{group } k
\end{aligned}
$$

and

$$h(t; \mathbf{x}) = h_0(t) \prod_{\ell=1}^{k} \psi_\ell^{x_\ell} = \begin{cases} h_0(t), & \mathbf{x} = (0, 0, \ldots, 0) \\ h_0(t)\psi_j, & x_j = 1, \quad j = 1, \ldots k \end{cases}$$

With $\psi_j = e^{\beta_j}$, $j = 1, \ldots, k$, we get

$$h(t; \mathbf{x}) = h_0(t)e^{x_1\beta_1 + x_2\beta_2 + \cdots + x_k\beta_k} = h_0(t)e^{\mathbf{x}\beta}, \tag{3.5}$$

where $\beta = (\beta_1, \beta_2, \ldots, \beta_k)$.

### 3.3.3 The general Cox regression model

We may now generalize equation (3.5) by letting the components of $\mathbf{x}_i$ take *any value*. Let data and model take the following form:

Data:

$$(t_{i0}, t_i, d_i, \mathbf{x}_i), \ i = 1, \ldots, n, \tag{3.6}$$

where $t_{i0}$ is the *left truncation time point* (if $t_{i0} = 0$ for all $i$, then this variable may be omitted), $t_i$ is the *end time point*, $d_i$ is the *"event indicator"* (1 or *TRUE* if event, else 0 or *FALSE*), and $\mathbf{x}_i$ is a vector of *explanatory variables*.

Model:

$$h(t; \mathbf{x}_i) = h_0(t)e^{\mathbf{x}_i\beta}, \quad t > 0. \tag{3.7}$$

This is a *regression model* where the *response variable* is $(t_0, t, d)$ (we will call it a *survival object*) and the *explanatory variable* is $\mathbf{x}$, possibly (often) vector valued.

In equation (3.7) there are two components to estimate, the regression coefficients $\beta$, and the *baseline hazard function* $h_0(t)$, $t > 0$. For the former task, the *partial likelihood* (Cox, 1975) is used. See Appendix A for a brief summary.

## 3.4   Estimation of the Baseline Cumulative Hazard Function

The usual estimator (continuous time) of the baseline cumulative hazard function is

$$\hat{H}_0(t) = \sum_{j:t_j \leq t} \frac{d_j}{\sum_{m \in R_j} e^{\mathbf{x}_m \hat{\beta}}}, \qquad (3.8)$$

where $d_j$ is the number of events at $t_j$ and $\hat{\beta}$ is the result of maximizing the partial likelihood. Note that if $\hat{\beta} = 0$, this reduces to

$$\hat{H}_0(t) = \sum_{j:t_j \leq t} \frac{d_j}{n_j}, \qquad (3.9)$$

the Nelson-Aalen estimator. In equation (3.9), $n_j$ is the size of $R_j$.

In the **R** package **eha**, the baseline hazard is estimated at the value zero of the covariates (but at the reference category for a factor covariate). This is different from practice in other packages, where some average value of covariate values are chosen as reference. However, this is generally a bad habit, because it will frequently lead to a variation in the reference value as subsets of the original data set is taken. Instead, the practitioner must exercise judgment in choosing relevant reference values for (continuous) covariates. There are two instances when these choices make a difference: (i) When interactions are present, main effect estimates will vary with choice (more about this later), and (ii) estimation of baseline hazards and survival functions.

In order to calculate the cumulative hazards function for an individual with a specific covariate vector $\mathbf{x}$, use the formula

$$\hat{H}(t; \mathbf{x}) = \hat{H}_0(t) e^{\mathbf{x} \hat{\beta}}.$$

The corresponding survival functions may be estimated by the relation

$$\hat{S}(t; \mathbf{x}) = \exp(-\hat{H}(t; \mathbf{x}))$$

It is also possible to use the terms in the sum equation (3.8) to build an estimator analogous to the Kaplan-Meier estimator equation (2.9). In practice, there is no big difference between the two methods. For more on interpretation of parameter estimates and model selection, see the appropriate chapter.

## 3.5 Proportional Hazards in Discrete Time

In discrete time, the hazard function is, as we saw earlier, a set of conditional probabilities, and so its range is restricted to the interval $(0, 1)$. Therefore, the definition of proportional hazards used for continuous time is unpractical; the multiplication of a probability by a constant may result in a quantity larger than one.

One way of motivating proportional hazards in discrete time is to regard the discreteness as a result of grouping true continuous time data, for which the proportional hazards assumption hold. For instance, in a follow-up study of human mortality, we may only have data recorded once a year, and so life length can only be measured in years. Thus, we assume that there is a true exact life length $T$, but we can only observe that it falls in an interval $(t_i, t_{i+1})$.

Assume *continuous* proportional hazards, and a *partition* of time:

$$0 = t_0 < t_1 < t_2 < \cdots < t_k = \infty.$$

Then
$$P(t_{j-1} \le T < t_j \mid T \ge t_{j-1}; \mathbf{x}) = \frac{S(t_{j-1} \mid \mathbf{x}) - S(t_j \mid \mathbf{x})}{S(t_{j-1} \mid \mathbf{x})}$$

$$= 1 - \frac{S(t_j \mid \mathbf{x})}{S(t_{j-1} \mid \mathbf{x})} = 1 - \left(\frac{S_0(t_j)}{S_0(t_{j-1})}\right)^{\exp(\beta\mathbf{x})}$$

$$= 1 - (1 - h_j)^{\exp(\beta\mathbf{x})} \quad (3.10)$$

with $h_j = P(t_{j-1} \le T < t_j \mid T \ge t_{j-1}; \mathbf{x} = \mathbf{0})$, $j = 1, \ldots, k$. We take equation (3.10) as the *definition* of proportional hazards in discrete time.

### 3.5.1  Logistic regression

It turns out that a proportional hazards model in discrete time, according to definition equation (3.10), is nothing else than a *logistic regression* model with the *cloglog link* (cloglog is short for "complementary log-log" or $\beta\mathbf{x} = \log(-\log(p))$). In order to see this, let

$$(1 - h_j) = \exp(-\exp(\alpha_j)), \ j = 1, \ldots, k, \quad (3.11)$$

and

$$X_j = \begin{cases} 1, & t_{j-1} \le T < t_j \\ 0, & \text{otherwise} \end{cases}, \quad j = 1, \ldots, k.$$

Then

$$P(X_1 = 1; \mathbf{x}) = 1 - \exp(-\exp(\alpha_1 + \beta\mathbf{x}))$$
$$P(X_j = 1 \mid X_1 = \cdots = X_{j-1} = 0; \mathbf{x}) = 1 - \exp(-\exp(\alpha_j + \beta\mathbf{x})),$$
$$j = 2, \ldots, k.$$

This is logistic regression with a cloglog link. Note that extra parameters $\alpha_1, \ldots, \alpha_k$ are introduced, one for each potential event time (interval). They correspond to the baseline hazard function in continuous time, and are be estimated simultaneously with the regression parameters.

## 3.6 Doing It in R

We utilize the *child mortality* data, Skellefteå 1850–1884, to illustrate some aspects of an ordinary Cox regression. Newborn in the parish are sampled between 1820 and 1840 and followed to death or reaching the age of 15, when they are right censored. So in effect child mortality in the ages 0–15 is studied. The data set is named child and available in eha.

There are two **R** functions that are handy for a quick look at a data frame, str and head. The function head prints the first few lines (observations) of a data frame (there is also a corresponding tail function that prints a few of the *last* rows).

```
library(eha) #Loads also the data frame 'mort'.
ch <- child[, c("birthdate", "sex", "socBranch",
                "enter", "exit", "event")] # Select columns.
head(ch)
```

|     | birthdate  | sex    | socBranch | enter | exit   | event |
| --- | ---------- | ------ | --------- | ----- | ------ | ----- |
| 3   | 1853-05-23 | male   | farming   | 0     | 15.000 | 0     |
| 42  | 1853-07-19 | male   | farming   | 0     | 15.000 | 0     |
| 47  | 1861-11-17 | male   | worker    | 0     | 15.000 | 0     |
| 54  | 1872-11-16 | male   | farming   | 0     | 15.000 | 0     |
| 78  | 1855-07-19 | female | worker    | 0     | 0.559  | 1     |
| 102 | 1855-09-29 | male   | farming   | 0     | 0.315  | 1     |

The function str gives a summary description of the *structure* of an **R** object, often a *data frame*

```
str(ch)
```

```
'data.frame':   26574 obs. of  6 variables:
 $ birthdate: Date, format: "1853-05-23" ...
 $ sex      : Factor w/ 2 levels "male","female": 1 1 1 1 2 1 1 1 2 2 ...
 $ socBranch: Factor w/ 4 levels "official","farming",..: 2 2 4 2 4 2 2 2 2 2 ...
 $ enter    : num  0 0 0 0 0 0 0 0 0 0 ...
 $ exit     : num  15 15 15 15 0.559 0.315 15 15 15 15 ...
 $ event    : num  0 0 0 0 1 1 0 0 0 0 ...
```

First, there is information about the data frame: It *is* a `data.frame`,
with six variables measured on 26574 objects. Then each variable
is individually described: name, type, and a few of the first values.
The values are usually rounded to a few digits. The *Factor* lines
are worth noticing: They describe of course `factor` covariates and
the levels. Internally, the levels are coded 1, 2, ..., respectively.

Also note the variable `birthdate`: It is of the `Date` type, and it has
some quirks when used in regression models.

```
res <- coxreg(Surv(exit, event) ~ sex + socBranch + birthdate,
              data = ch)
print(summary(res), digits = 4)
```

| Covariate | | Mean | Coef | Rel.Risk | S.E. | LR p |
|---|---|---|---|---|---|---|
| sex | | | | | | 0.0019 |
| | male | 0.510 | 0 | 1 (reference) | | |
| | female | 0.490 | -0.083 | 0.920 | 0.027 | |
| socBranch | | | | | | 0.0001 |
| | official | 0.021 | 0 | 1 (reference) | | |
| | farming | 0.710 | -0.017 | 0.983 | 0.092 | |
| | business | 0.011 | 0.330 | 1.391 | 0.141 | |
| | worker | 0.258 | 0.099 | 1.104 | 0.094 | |
| birthdate | | 1869-07-13 | -0.000 | 1.000 | 0.000 | 0.0000 |

| | |
|---|---|
| Events | 5616 |
| Total time at risk | 325030 |

```
Max. log. likelihood        -56481
LR test statistic           67.10
Degrees of freedom          5
Overall p-value             4.11227e-13
```

Note that the coefficient for birthdate is equal to 0.0000 (with four decimals), yet it is significantly different from zero. The explanation is that the time unit behind Dates is *day*, and it would be more reasonable to use *year* as time unit. It can be accomplished by creating a new covariate, call it cohort, which is *birth year*. It is done with the aid of the functions toTime (in eha) and floor:

```
ch$cohort <- floor(toTime(ch$birthdate))
fit <- coxreg(Surv(exit, event) ~ sex + socBranch + cohort,
              data = ch)
print(summary(fit), digits = 4, short = TRUE)
```

| Covariate | | Mean | Coef | Rel.Risk | S.E. | LR p |
|---|---|---|---|---|---|---|
| sex | | | | | | 0.0018 |
| | male | 0.510 | 0 | 1 (reference) | | |
| | female | 0.490 | -0.083 | 0.920 | 0.027 | |
| socBranch | | | | | | 0.0001 |
| | official | 0.021 | 0 | 1 (reference) | | |
| | farming | 0.710 | -0.017 | 0.984 | 0.092 | |
| | business | 0.011 | 0.330 | 1.390 | 0.141 | |
| | worker | 0.258 | 0.099 | 1.104 | 0.094 | |
| cohort | | 1869.035 | -0.008 | 0.992 | 0.001 | 0.0000 |

Note the argument short = TRUE in the print statement. It suppresses the general statistics that were printed in the previous output below the table of coefficients.

### 3.6.1 The estimated baseline cumulative hazard function

The estimated baseline cumulative function (see equation (3.8)) is preferably reported by a graph using the plot command, see Figure 3.8.

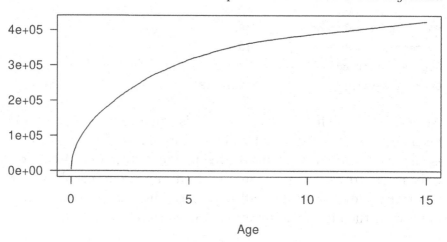

**FIGURE 3.8** Estimated cumulative baseline hazard function.

The figure shows the baseline cumulative hazard function for an individual with the value zero on all covariates. What does it mean here? There are two covariates, the first is ses with reference value lower, and the second is birthdate, with reference value 0! So Figure 3.8 shows the estimated cumulative hazards function for a man born on January 1 in the year 0 (which does not exist, by the way) and who belongs to a lower social class. This is not a reasonable extrapolation, and we need to get a new reference value for birthdate. It should be within the range of the observed values, which is given by the range function:

```
range(ch$cohort)
```

```
[1] 1850 1884
```

So a reasonable value to subtract is 1860 (or any meaningful value in the range):

```
ch$cohort <- ch$cohort - 1860
res3 <- coxreg(Surv(exit, event) ~ sex + socBranch + cohort,
          data = ch)
print(summary(res3), digits = 4, short = TRUE)
```

| Covariate | | Mean | Coef | Rel.Risk | S.E. | LR p |
|---|---|---|---|---|---|---|
| sex | | | | | | 0.0018 |
| | male | 0.510 | 0 | 1 (reference) | | |
| | female | 0.490 | -0.083 | 0.920 | 0.027 | |
| socBranch | | | | | | 0.0001 |
| | official | 0.021 | 0 | 1 (reference) | | |
| | farming | 0.710 | -0.017 | 0.984 | 0.092 | |
| | business | 0.011 | 0.330 | 1.390 | 0.141 | |
| | worker | 0.258 | 0.099 | 1.104 | 0.094 | |
| cohort | | 9.035 | -0.008 | 0.992 | 0.001 | 0.0000 |

As you can see, nothing changes regarding the parameter estimates
(because no interactions are present). What about the baseline
cumulative hazards graph?

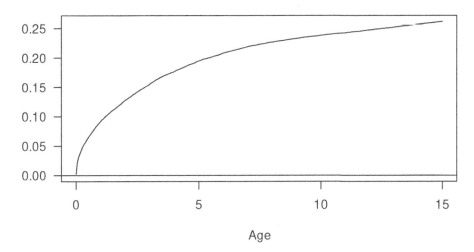

**FIGURE 3.9** Estimated cumulative baseline hazard function,
centered birthdate at 1810.

The two Figures 3.8 and 3.9 are identical, *except* for the numbers on the $y$ axis. This would *not* hold for the corresponding plots of the baseline *survival* function (try it!).

# 4

## Explanatory Variables and Regression

This chapter deals with the various forms the explanatory variables can take and their possible interplay. In principle, this applies to all forms of regression analysis, but here we concentrate on survival models.

Explanatory variables, or *covariates*, may be of essentially two different types, *continuous* and *discrete*. The discrete type usually takes only a finite number of distinct values and is called a `factor` in **R**. A special case of a factor is one that takes only two distinct values, say 0 and 1. Such a factor is called an *indicator*, because we can let the value 1 indicate the presence of a certain property and 0 denote its absence. To summarize, there is

- **Covariate**: taking values in an *interval* (*age, blood pressure, weight*).
- **Factor**: taking a *finite* number of values (*civil status, occupation, cohort*).
- **Indicator**: a factor taking *two* values (*sex, born in Umeå*).

Usually, we do not distinguish between factors and indicators, they are all factors.

## 4.1 Continuous Covariates

We use the qualifier *continuous* to stress that factors are excluded, because often the term *covariate* is used as a synonym for *explanatory variable*.

DOI: 10.1201/9780429503764-4

Values taken by a continuous covariate are *ordered*. The *effect* on the response is by model definition ordered in the *same* or *reverse* order. On the other hand, values taken by a factor are *unordered* (but may be defined as ordered in **R**).

## 4.2   Factor Covariates

An explanatory variable that can take only a finite (usually small) number of distinct values is called a *categorical variable*. In **R** language, it is called a *factor*. Examples of such variables are *gender, socio-economic status*, and *birth place*. Students of statistics have long been taught to create *dummy variables* in such situations, in the following way:

- Given a categorical variable $F$ with $(k+1)$ levels $(f_0, f_1, f_2, \ldots f_k)$ ($k+1$ levels),
- Create $k$ *indicator* ("dummy") variables $(I_1, I_2, \ldots I_k)$.

The level $f_0$ is the reference category, characterized by that all indicator variables are zero for an individual with this value. Generally, for the level, $f_i$, $i = 1, \ldots, k$, the indicator variable $I_i$ is one, the rest are zero. In other words, for a single individual, at most one indicator is one, and the rest are zero.

In **R**, there is no need to explicitly create dummy variables, it is done behind the scenes by the functions `factor` and `as.factor`.

Note that a factor with *two* levels, i.e., an indicator variable, can always be treated as a continuous covariate, if coded numerically (e.g., 0 and 1).

**Example 4.1** (Infant mortality and age of mother)

Consider a demographic example, the influence of mother's age (a continuous covariate) on infant mortality. It is considered well-known that a *young* mother means high risk for the infant, and also that *old* mother means high risk, compared to *"in-between-aged"* mothers. So the risk order is not the same (or reverse) as the age order.

One solution (not necessarily the best) to this problem is to *factorize*: Let, for instance,

$$\text{mother's age} = \begin{cases} \text{low,} & 15 < \text{age} \le 25 \\ \text{middle,} & 25 < \text{age} \le 35 \\ \text{high,} & 35 < \text{age} \le 50 \end{cases}$$

In this layout, there will be two parameters measuring the deviation from the reference category, which will be the first category by default.

In **R**, this is easily achieved with the aid of the cut function. It works like this, illustrating it with the data set child:

```
age.group <- cut(child$m.age, c(15, 25, 35, 51))
table(age.group, useNA = "ifany")


age.group
(15,25] (25,35] (35,51]
   3917   13805    8852
```

Note that the created intervals by default are closed to the right and open to the left. This has consequences for how observations exactly on a boundary are treated; they belong to the lower-valued interval. The argument right in the call to cut can be used to switch this behavior the other way around.

Note further that values falling below the smallest value (15 in our example) or above the largest value (51) are reported as *missing values* (NA in **R** terminology, Not Available). For further information about the use of the cut function, see its help page.

An infant mortality analysis with the created factor covariate may look like this:

```
ch <- child
ch$age.group <- age.group
ch <- age.window(ch, c(0, 1)) # Right censor at age one.
fit <- coxreg(Surv(enter, exit, event) ~ age.group, data = ch)
print(summary(fit), short = TRUE)
```

| Covariate | Mean | Coef | Rel.Risk | S.E. | LR p |
|---|---|---|---|---|---|
| age.group | | | | | 0.008 |
| (15,25] | 0.147 | 0 | 1 (reference) | | |
| (25,35] | 0.522 | -0.056 | 0.945 | 0.064 | |
| (35,51] | 0.332 | 0.090 | 1.094 | 0.067 | |

Obviously, it is safest for the infants whose mothers' age lies in the span 25–35 years of age. □

## 4.3  Interactions

The meaning of *interaction* between two explanatory variables is described by an example, walking through the three possible combinations of covariate types.

**Example 4.2** (Old age mortality)

We return to the oldmort data set, slightly modified, see Table 4.1, where the first five of a total to 6495 rows are shown.

The sampling frame is a rectangle in the Lexis diagram, see Figure 4.1. It can be described as all persons born between January 1, 1775 and January 1, 1821, and present in the solid rectangle at some moment.

There are four possible explanatory variables behind survival after age 60 in the example.

**TABLE 4.1** Old age mortality, Sundsvall 1860–1880

| birthdate | enter | exit | event | sex | farmer | imr.birth |
|---|---|---|---|---|---|---|
| 1775.286 | 84.714 | 85 | 0 | female | no | 14.20722 |
| 1775.092 | 84.908 | 85 | 0 | female | yes | 11.97183 |
| 1775.374 | 84.626 | 85 | 0 | female | no | 16.93548 |
| 1775.859 | 84.141 | 85 | 0 | male | yes | 16.93548 |
| 1775.111 | 84.889 | 85 | 0 | female | no | 12.70903 |

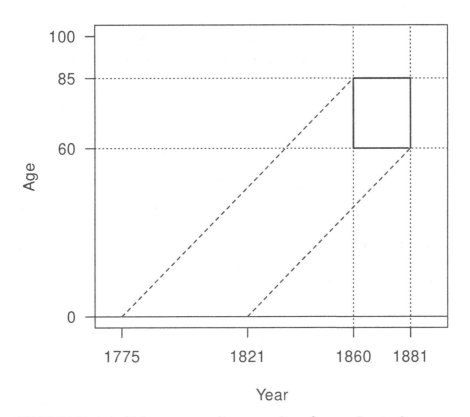

**FIGURE 4.1** Old age mortality sampling frame, Lexis diagram.

$$\text{farmer} = \begin{cases} \text{no} \\ \text{yes} \ (e^\alpha) \end{cases}$$

$$\text{sex} = \begin{cases} \text{male} \\ \text{female} \ (e^\beta) \end{cases}$$

$$\text{birthdate} \ (x_1) \qquad\qquad (e^{\gamma_1 x_1})$$

$$\text{IMR at birth} \ (x_2) \qquad\qquad (e^{\gamma_2 x_2})$$

The two first are factors, and the last two are continuous covariates. We go through the three distinct combinations and illustrate the meaning of interaction in each. □

### 4.3.1 Two factors

Assume that we, in the oldmort example, only have the factors sex and farmer at our disposal as explanatory variables. We may fit a Cox regression model like this:

```
fit <- coxreg(Surv(enter, exit, event) ~ sex + farmer,
          data = om)
```

The output from coxreg (the *model fit*) is saved in the object fit, which is investigated, first by the function summary, which here performs successive $\chi^2$ tests of the significance of the estimated regression coefficients (the null hypothesis in each case is that the true coefficient is zero).

```
print(summary(fit), short = TRUE)
```

| Covariate | Mean | Coef | Rel.Risk | S.E. | LR p |
|---|---|---|---|---|---|
| sex | | | | | 0.000 |
| male | 0.407 | 0 | 1 (reference) | | |
| female | 0.593 | -0.232 | 0.793 | 0.048 | |

| farmer | | | | 0.025 |
|---|---|---|---|---|
| no | 0.772 | 0 | 1 (reference) | |
| yes | 0.228 | -0.127 | 0.880 | 0.058 |

The $p$-value corresponding to the coefficient for sex is *very* small, meaning that there is a highly significant difference in mortality between women and men. The difference in mortality between farmers and non-farmers is also statistically significant, but with a larger $p$-value. Next, we want to see the *size* of the difference in mortality between women and men, and between farmers and non-farmers, so look at the parameter estimates, especially in the column "Rel. Risk". It tells us that female mortality is 79.3 per cent of the male mortality, and that the farmer mortality is 88.0 per cent of the non-farmer mortality. Furthermore, this model is *multiplicative* (additive on the log scale), so we can conclude that the mortality of a female farmer is $0.793 \times 0.880 \times 100 = 69.8$ per cent of that of a male non-farmer.

We can also illustrate the sex difference graphically by a stratified analysis, see Figure 4.2.

```
par(las = 1)
fit2 <- coxreg(Surv(enter, exit, event) ~ strata(sex) + farmer,
          data = om)
plot(fit2, fun = "cumhaz", xlim = c(60, 85),
     lty = 1:2, xlab = "Age")
abline(h = 0)
```

Obviously the proportionality assumption is well met.

An obvious question to ask is: Is the sex difference in mortality the same among farmers as among non-farmers? In other words: Is there an *interaction* between sex and farmer/non-farmer regarding the effect on mortality?

In order to test for an interaction in **R**, the plus (+) sign in the formula is changed to a multiplication sign (∗):

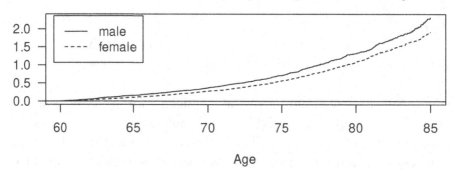

**FIGURE 4.2** Cumulative hazards for men and women.

```
fit4 <- coxreg(Surv(enter, exit, event) ~ sex * farmer,
          data = om)
print(x <- summary(fit4), short = TRUE)
```

```
Single term deletions

Model:
Surv(enter, exit, event) ~ sex * farmer
           Df   AIC LRT Pr(>Chi)
<none>          25912
sex:farmer  1 25916 6.4     0.011
```

| Covariate | Mean | Coef | Rel.Risk | S.E. | Wald p |
|---|---|---|---|---|---|
| sex | | | | | |
| male | 0.407 | 0 | 1 (reference) | | |
| female | 0.593 | -0.297 | 0.743 | 0.054 | 0.000 |
| farmer | | | | | |
| no | 0.772 | 0 | 1 (reference) | | |
| yes | 0.228 | -0.252 | 0.777 | 0.076 | 0.001 |
| sex:farmer | | | | | |
| female:yes | | 0.291 | 1.338 | 0.114 | 0.011 |

Note first that the output looks different when interactions are present: The first part is a chisquare test of the significance of the interaction(s), then the usual table with parameter coefficients are presented with one difference: The Wald, and not the LR $p$-values are given. The small LR $p$-value (1.1 per cent) indicates that there is a *statistically* significant (at the 5 per cent level) interaction effect. The *size* of the interaction effect is seen in the table of estimated coefficients. Note that the reference is a *non-farmer man*, so the "main" effect for sex == female is in fact comparing a non-farming woman to a non-farming man with the result that the woman has a mortality that is 74.3 per cent of that of a corresponding man. When comparing a farming woman to a farming man, we have to take the interaction into account, and the result is that a farming woman has a risk that is $0.743 \times 1.338 \times 100 = 99.4$ per cent of the risk for the farming man. Conclusion: Among farmers, the sex difference in mortality is negligible, but in the rest of the population it is large. How the *causality* works here is of course a completely different matter. We may guess that farmers are often married, for instance, we do not find any unmarried female farmers in our data.

In cases like this, it is often best to perform separate analyses for women and men, or, alternatively, for farmers and non-farmers. Do that as an exercise and interpret your results.

### 4.3.2   One factor and one continuous covariate

Now the covariates sex (factor) and birthdate (continuous) are in focus. First the model without interaction is fitted.

```
fit5 <- coxreg(Surv(enter, exit, event) ~ sex + birthdate,
            data = om)
print(summary(fit5), short = TRUE)
```

| Covariate | | Mean | Coef | Rel.Risk | S.E. | LR p |
|---|---|---|---|---|---|---|
| sex | | | | | | 0.000 |
| | male | 0.407 | 0 | 1 (reference) | | |
| | female | 0.593 | -0.207 | 0.813 | 0.047 | |
| birthdate | | 1802.908 | -0.006 | 0.994 | 0.004 | 0.135 |

Obviously sex is statistically significant, while birthdate is not. We go directly to the estimated coefficients after adding an interaction term to the model:

```
fit6 <- coxreg(Surv(enter, exit, event) ~ sex * birthdate,
               data = om)
round(summary(fit6)$coefficients[, 1:2], 5)
```

```
                        coef  exp(coef)
sexfemale            9.12285 9162.24554
birthdate           -0.00327    0.99673
sexfemale:birthdate -0.00519    0.99483
```

What is going on here? The main effect of sex is huge! The answer lies in the interpretation of the coefficients: (i) To evaluate the effect on mortality for an individual with given values of the covariates, we must involve the interaction coefficient. For instance, a female born on January 2, 1801 (birthdate = 1801.003) will have the relative risk

$$\exp(9.12285 - 0.00327 \times 1801.003 - 0.00519 \times 1801.003) = 0.00185$$

compared to a man with birthdate= 0, the reference values! Obviously, it is ridiculous to extrapolate the model 1800 years back in time, but this is nevertheless the correct way to interpret the coefficients. So, at birthdate $= 0$, women's mortality is 9162 times higher than that of men! (ii) With interactions involved, it is strongly recommended to *center* continuous covariates. In our example, we could do that by subtracting 1810 (say) from birthdate. That makes sex = "male", birthdate = 1800 the new *reference values*. The result is

```
                       coef exp(coef)
sexfemale           -0.26429   0.76775
birthdate           -0.00327   0.99673
sexfemale:birthdate -0.00519   0.99483
```

This is more reasonable-looking: At the beginning of the year 1810, woman had a mortality of 76.8 per cent of that of men.

To summarize: The effects can be illustrated by two curves, one for men and one for women. The coefficients for sex, 0 and −0.26429, are the respective intercepts, and −0.00327 and −0.00846 are the slopes. See Figure 4.3 for a graphical illustration.

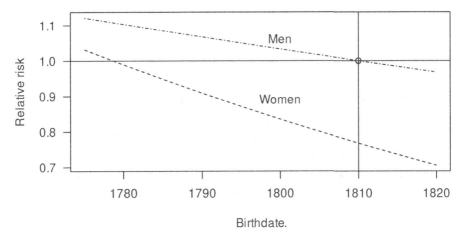

**FIGURE 4.3** The effect of birthdate on relative mortality for men and women. Reference: Men, birthdate = 1810.

### 4.3.3 Two continuous covariates

In our example data, the covariates birthdate and imr.birth are continuous, and we get

```
fit8 <- coxreg(Surv(enter, exit, event) ~ birthdate * imr.birth,
            data = om)
(res <- round(summary(fit8)$coefficients[, 1:2], 5))
```

|  | coef | exp(coef) |
|---|---|---|
| birthdate | −0.03024 | 0.97021 |
| imr.birth | 0.02004 | 1.02024 |
| birthdate:imr.birth | 0.00158 | 1.00158 |

The interpretation of the coefficients are: The reference point is `birthdate = 0` (1810) and `imr.birth = 0`, and if `birthdate = x` and `imr.birth = y`, then the relative risk is

$$\exp(-0.03024x + 0.02004y + 0.00158xy),$$

analogous to the calculation in the case with one continuous and one factor covariate.

---

## 4.4   Interpretation of Parameter Estimates

In the proportional hazards model, the parameter estimates are logarithms of risk ratios relative to the baseline hazard. The precise interpretations of the coefficients for the two types of covariates are discussed. The conclusion is that $e^{\beta}$ has a more direct and intuitive interpretation than $\beta$ itself.

### 4.4.1   Continuous covariate

If $x$ is a continuous covariate, and $h(t; x) = h_0(t)e^{\beta x}$, then

$$\frac{h(t; x+1)}{h(t; x)} = \frac{h_0(t)e^{\beta(x+1)}}{h_0(t)e^{\beta x}} = e^{\beta}.$$

so the risk increases with a factor $e^{\beta}$, when $x$ is increased by one unit. In other words, $e^{\beta}$ is a *relative risk* (or a *hazard ratio*, which often is a preferred term in certain professions).

### 4.4.2   Factor

For a factor covariate, in the usual coding with a reference category, $e^{\beta}$ is the relative risk compared to that reference category.

## 4.5 Model Selection

In regression modeling, there is often several competing models for describing data. In general, there are no strict rules for "correct selection". However, for *nested* models, there are some formal guidelines. For a precise definition of this concept, see Appendix A.

### 4.5.1 Model selection in general

Some general advise regarding model selection is given here.

- Remember, there are no *true* models, only some *useful* ones. This statement is attributed to G.E.P. Box.
- More than one model may be useful.
- Keep *important* covariates in the model.
- Avoid automatic stepwise procedures!
- If interaction terms are present, the corresponding main terms must be there. (There are some exceptions to this rule, see for instance the example about infant and maternal mortality in Chapter 10.)

# 5

## Poisson Regression

Here *Poisson regression* is introduced and its connection to Cox regression is discussed. We start by defining the Poisson distribution, then discuss the connection to Cox regression and tabular lifetime data.

### 5.1 The Poisson Distribution

The *Poisson distribution* is used for *count data*, i.e., when the result may be any positive integer *0, 1, 2, ...*, without upper limit. The *probability density function* (pdf) *P* of a random variable $X$ following a Poisson distribution is

$$P(X = k) = \frac{\lambda^k}{k!} \exp(-\lambda), \quad \lambda > 0; \; k = 0, 1, 2, ..., \qquad (5.1)$$

The parameter $\lambda$ is both the mean and the variance of the distribution. In Figure 5.1 the pdf (5.1) is plotted for some values of $\lambda$.

Note that when $\lambda$ increases, the distribution looks more and more like a normal distribution.

In **R**, the Poisson distribution is represented by four functions, dpois ppois, qpois, and rpois, representing the probability density function (pdf), the cumulative distribution function (cdf), the quantile function (the inverse of the cdf), and random number generation, respectively. See the help page for the Poisson distribution for more detail. In fact, this is the scheme present for all probability distributions available in **R**.

DOI: 10.1201/9780429503764-5

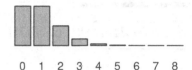

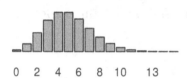

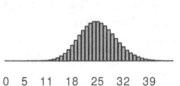

**FIGURE 5.1** The Poisson cdf for different values of the mean.

For example, the upper left bar plot in Figure 5.1 is produced in **R** by

```
barplot(dpois(0:5, lambda = 0.5), axes = FALSE,
        main = expression(paste(lambda, " = ", 0.5)))
```

Note that the function dpois is *vectorizing*:

```
round(dpois(0:5, lambda = 0.5), 5)
```

```
[1] 0.60653 0.30327 0.07582 0.01264 0.00158 0.00016
```

**Example 5.1** (Marital fertility)

As an example where the Poisson distribution may be relevant, we look at the number of children born to a woman after marriage. The data frame *fert* in eha can be used to calculate the number of births per married woman in Skellefteå during the 19th century; however, this data set contains only marriages with one or more births. Let us instead count the number of births beyond one. The result is shown in Figure 5.2.

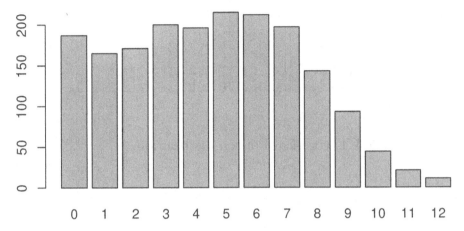

**FIGURE 5.2** Number of children beyond one for married women with at lest one child.

The question is: Does this look like a Poisson distribution? One way of checking this is to plot the *theoretic* distribution with the same mean (the parameter $\lambda$) as the sample mean in the data.

```
lam <- mean(kids)
barplot(dpois(0:12, lambda = lam))
```

The result is shown in Figure 5.3.

Obviously, the fertility data do not follow the Poisson distribution so well. It is in fact *over-dispersed* compared to the Poisson distribution. A simple way to check that is to calculate the sample mean and variance of the data. If data come from a Poisson distribution, these numbers should be equal (theoretically) or reasonably close.

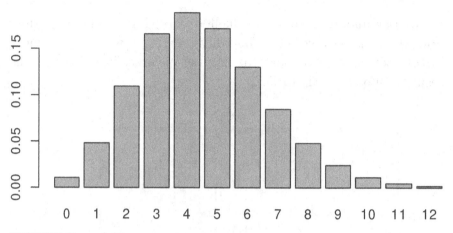

**FIGURE 5.3** Theoretical Poisson distribution.

```
mean(kids)
```

```
[1] 4.548682
```

```
var(kids)
```

```
[1] 8.586838
```

They are not very close, which also is obvious from comparing the graphs. □

## 5.2   The Connection to Cox Regression

There is an interesting connection between Cox regression and Poisson regression, which can be illustrated as follows.

```
dat <- data.frame(enter = rep(0, 4), exit = 1:4,
                  event = rep(1, 4), x = c(0, 1, 0, 1))
dat
```

```
  enter exit event x
1     0    1     1 0
2     0    2     1 1
3     0    3     1 0
4     0    4     1 1
```

We have generated a very simple data set, dat. It consists of four life lengths with a covariate $x$. A Cox regression, where the covariate $x$ is related to survival is given by

```
library(eha)
fit <- coxreg(Surv(enter, exit, event) ~ x, data = dat)
print(summary(fit), short = TRUE)
```

| Covariate | Mean | Coef | Rel.Risk | S.E. | LR p |
|-----------|------|------|----------|------|------|
| x | 0.600 | -0.941 | 0.390 | 1.240 | 0.432 |

The baseline hazards is given by the function hazards. By default it calculates the *cumulative* hazards, but here we need the non-cumulative ones, which is accomplished by specifying cum = FALSE:

```
haz <- hazards(fit, cum = FALSE)
haz
```

```
$`1`
     [,1]      [,2]
[1,]    1 0.3596118
```

```
[2,]     2 0.5615528
[3,]     3 0.7192236
[4,]     4 2.5615528
```

```
attr(,"class")
[1] "hazdata"
```

Note that haz is a *list* with one component per stratum. In this case there is only one stratum, so the list has one component. This component is a matrix with two columns; the first contains the observed distinct failure times, and the second the corresponding estimates of the hazard "atoms".

Now, this can be replicated by Poisson regression! First the data set is summarized around each observed failure time. That is, a snapshot of the risk set at each failure time is created by the function toBinary.

```
library(eha)
datB <- toBinary(dat)
datB
```

|      | event | riskset | risktime | x | orig.row |
|------|-------|---------|----------|---|----------|
| 1    | 1     | 1       | 1        | 0 | 1        |
| 2    | 0     | 1       | 1        | 1 | 2        |
| 3    | 0     | 1       | 1        | 0 | 3        |
| 4    | 0     | 1       | 1        | 1 | 4        |
| 2.1  | 1     | 2       | 2        | 1 | 2        |
| 3.1  | 0     | 2       | 2        | 0 | 3        |
| 4.1  | 0     | 2       | 2        | 1 | 4        |
| 3.2  | 1     | 3       | 3        | 0 | 3        |
| 4.2  | 0     | 3       | 3        | 1 | 4        |
| 4.3  | 1     | 4       | 4        | 1 | 4        |

Note that three new "covariates" are created, riskset, risktime, and orig.row. In the Poisson regression to follow, the variable riskset must be included as a *factor*. The Poisson regression is performed with the glm function:

```
fit2 <- glm(event ~ riskset + x, family = poisson,
            data = datB)
(co <- coefficients(fit2))
```

```
(Intercept)     riskset2     riskset3     riskset4            x
 -1.0227302    0.4456807    0.6931472    1.9633438   -0.9406136
```

```
co[2:4] <- co[2:4] + co[1] # Calculate the hazard atoms.
haz.glm <- exp(co[1:4])
```

The parameter estimate corresponding to $x$ is exactly the same as in the Cox regression. The baseline hazard function is estimated with the aid of the *(Intercept)* and the *riskset* estimates, and we compare them below with the hazards estimates from coxreg.

```
xx <- cbind(haz[[1]], haz.glm)
colnames(xx) <- c("Time", "coxreg", "glm")
xx
```

|             | Time | coxreg    | glm       |
|-------------|------|-----------|-----------|
| (Intercept) | 1    | 0.3596118 | 0.3596118 |
| riskset2    | 2    | 0.5615528 | 0.5615528 |
| riskset3    | 3    | 0.7192236 | 0.7192236 |
| riskset4    | 4    | 2.5615528 | 2.5615528 |

The results are identical. This correspondence between Poisson and Cox regression was pointed out by Johansen (1983). However, this approach is not very useful in practical work with medium or large data sets, because the output from toBinary may be huge. For instance, with the data set child, the resulting data frame consists of slightly more than 56 million rows and 11 columns.

**Note:** There is a similar connection to *Binomial regression*: With
the method = "ml", we can compare output from coxreg and glm with
family = binomial.

```
fit.ml <- coxreg(Surv(enter, exit, event) ~ x,
                 method = "ml", data = dat)
fit.b <- glm(event ~ riskset + x,
             family = binomial(link = "cloglog"),
             data = datB)
rbind(coef(fit.ml),  coef(fit.b)["x"])
```

```
            x
[1,] -1.267101
[2,] -1.267101
```

Identical regression coefficients (and slightly different from what
what we got from the Poisson approach). More about discrete time
models in Chapter 7, "Parametric models".

## 5.3   The Connection to the Piecewise Constant Hazard Model

For completeness, the connection between the Poisson distribution
and the *piecewise constant hazard model* is mentioned here. This
is explored in detail in Chapter 7, where the Poisson formulation
is cast into a survival analysis costume and very straightforward
to use.

**TABLE 5.1** Swedish population data 2019–2020, ages 61–80.

| year | age | sex | deaths | pop |
|------|-----|-----|--------|-----|
| 2019 | 61 | women | 265 | 57152.0 |
| 2019 | 61 | men | 355 | 57393.0 |
| 2019 | 62 | women | 277 | 57064.5 |
| 2019 | 62 | men | 444 | 57430.0 |
| 2019 | 63 | women | 299 | 56436.0 |
| 2020 | 88 | men | 1757 | 11086.0 |
| 2020 | 89 | women | 2212 | 16405.0 |
| 2020 | 89 | men | 1662 | 9384.0 |
| 2020 | 90 | women | 2094 | 14229.0 |
| 2020 | 90 | men | 1614 | 7558.5 |

## 5.4  Tabular Lifetime Data

Although it is possible to use Poisson regression in place of Cox regression, the most useful application of Poisson regression in survival analysis is with tabular data. We have already seen an example of this in Chapter 2, the life and its connection to the survival function.

**Example 5.2** (Mortality in ages 61–90, Sweden 2019-2020.)

We create a data set swe (a subset and combination of swepop and swedeaths in eha), which contains population size and number of deaths by age and sex in Sweden for 2019 and 2020. The age span is restricted to ages 61–80. One reason for these choices is to get a grip on the effect of the covid-19 pandemic on the mortality increase in 2020 compared to 2019, pre-corona. Normally, mortality is lowered in all age groups from one year to the next, but 2019 to 2020 is different, see Table 5.1.

The Poisson model is, where $D_{ij}$ is number of deaths and $P_{ij}$ population size for age $i$ and sex $j$, $i = 61, \ldots, 80$; $j =$ female $(0)$, male $(1)$, and $\lambda_{ij}$ is the corresponding *mortality*,

$$D_{ij} \sim \text{Poisson}(\lambda_{ij} P_{ij}), \quad i = 61, 62, \ldots, 80; \ j = 0, 1,$$

with

$$\begin{aligned}\lambda_{61,j} P_{61,j} &= P_{61,j} \exp(\gamma + \beta * j) \\ &= \exp(\log(P_{61,j}) + \gamma + \beta * j), \quad j = 0, 1,\end{aligned}$$

and

$$\begin{aligned}\lambda_{ij} P_{ij} &= P_{ij} \exp(\gamma + \alpha_i + \beta * j) \\ &= \exp(\log(P_{ij}) + \gamma + \alpha_i + \beta * j), i = 62, 63, \ldots, 80; \ j = 0, 1.\end{aligned}$$

This calculation shows that it is the log of the population sizes, $\log(P_{ij})$, that is the correct *offset* to use in the Poisson regression. First we want age to be a factor (no restrictions like linearity), then the **R** function glm ("generalized linear model") is used to fit a Poisson regression model.

```
swe$age <- factor(swe$age)
fit <- glm(deaths ~ offset(log(pop)) + year + sex + age,
           family = poisson, data = swe)
drop1(fit, test = "Chisq")
```

```
Single term deletions

Model:
deaths ~ offset(log(pop)) + year + sex + age
        Df Deviance    AIC     LRT  Pr(>Chi)
<none>            152   1260
year     1        387   1493     235 < 2.2e-16
sex      1       4653   5760    4501 < 2.2e-16
age     29     117129 118180  116977 < 2.2e-16
```

The function drop1 is used above to perform two likelihood ratio tests. The results are that the three variables are all highly statistically significant, meaning that they are very important in describing the variation in mortality over year, sex, and age. To know *how*, we present (some of) the parameter estimates.

```
round(summary(fit)$coefficients[c(1:3), ], 3) # First 3 rows.
```

```
            Estimate Std. Error  z value Pr(>|z|)
(Intercept)   -5.453      0.029 -190.722        0
year2020       0.084      0.005   15.316        0
sexmen         0.370      0.006   66.995        0
```

```
round(summary(fit)$coefficients[c(19:21), ], 3) # Last three rows.
```

```
      Estimate Std. Error z value Pr(>|z|)
age77    1.685      0.032  53.329        0
age78    1.791      0.032  56.560        0
age79    1.945      0.031  61.788        0
```

The results so far tell us that males have a distinctly higher mortality than females, and that mortality steadily increases with age, no surprises so far. More interesting is to see that the covid-19 year 2020 has a significantly higher mortality than the year 2019. The parameter estimates for age also opens up the possibility to simplify the model by assuming a linear or quadratic effect of age on mortality. We leave that possibility for later.

One question remains: Is the 2019 advantage relatively the same over ages? To answer that question, we introduce an interaction:

```
fit1 <- glm(deaths ~ offset(log(pop)) + sex + year * age,
            family = poisson, data = swe)
drop1(fit1, test = "Chisq")
```

```
Single term deletions

Model:
deaths ~ offset(log(pop)) + sex + year * age
          Df Deviance    AIC    LRT Pr(>Chi)
<none>            107.2 1273.6
sex        1    4606.8 5771.2 4499.6  < 2e-16
year:age 29     152.1 1260.4   44.9  0.03019
```

Note that the drop1 function only tests the interaction term. This is because, as we saw before, the main effects tend to be meaningless in the presence of interaction. However, here there is no sign of interaction, and we can conclude that in a survival model with sex as covariate, we have *proportional hazards*! This is most easily seen in *graphs*. The first plot is based on the estimated Poisson model without interaction. Some calculations are first needed.

```
beta <- coefficients(fit)[2:3]
alpha <- coefficients(fit)[-(2:3)] # Remove sex and year
alpha[2:length(alpha)] <- alpha[2:length(alpha)] + alpha[1]
lambda.2019 <- exp(alpha)
lambda.2020 <- exp(alpha + beta[1])
```

Then the plot of the hazard functions by

```
par(las = 1)
plot(ages, lambda.2019, ylim = c(0, 0.15), type = "S",
    xlab = "Age", ylab = "Mortality")
lines(ages, lambda.2020, type = "S", lty = 2)
abline(h = 0)
legend("topleft", legend = c(2019, 2020), lty = 1:2)
```

Note the parameter ylim. It sets the scale on the $y$ axis, and some trial and error may be necessary to get a good choice. The function line adds a curve to an already existing plot, and the function abline adds a straight line; the argument h = 0 results in a *horizontal* line. See the help pages for these functions for further information.

The result is shown in Figure 5.4.

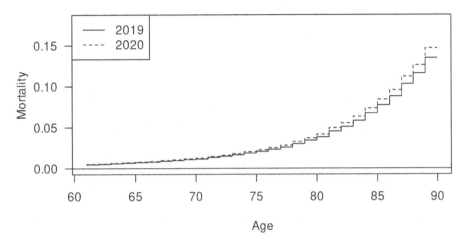

**FIGURE 5.4** Model based hazard functions for 2019 and 2020.

We can also make a plot based only on raw data. For that it is only necessary to calculate the *occurrence-exposure rates*. An occurrence-exposure rate is simply the ratio between the number of occurrences (of some event) divided by total exposure (waiting) time. In the present example, mean population size during one year is a very good approximation of the exact total exposure time (in years).

```
y2019 <- swe[swe$year == 2019, ]
y2019 <- aggregate(y2019[, c("deaths", "pop")],
                by = y2019["age"], FUN = sum)
y2020 <- swe[swe$year == 2020, ]
y2020 <- aggregate(y2020[, c("deaths", "pop")],
                by = y2020["age"], FUN = sum)
rate19 <- y2019$deaths / y2019$pop
rate20 <- y2020$deaths / y2020$pop
```

And finally the plot of the raw death rates is created as follows,

```
par(las = 1)
plot(ages, rate19, ylim = c(0, 0.045), xlab = "Age",
     ylab = "Mortality", type = "S")
lines(ages, rate20, lty = 2, type = "S")
abline(h = 0)
```

with the result as shown in Figure 5.5.

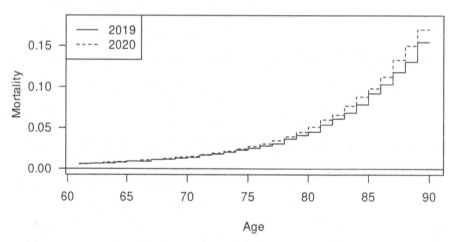

**FIGURE 5.5** Hazard functions for 2019 and 2020 based on raw data.

The differences between Figure 5.4 and Figure 5.5 are (i) the model-based hazard curves are higher than the raw ones, which depends on the fact that the model has *females* as reference category, and thus this figure depicts conditions for females, and (ii) we see small deviations from proportionality; the relative difference seems to increase with age. However, as the likelihood ratio implied, this finding is not statistically significant on the 5 per cent level.

A final remark: It was quite tedious, although straightforward, to apply raw Poisson regression to this problem. In Chapter 7, it will be shown how to use the function tpchreg in the **eha** package for a smooth and simple way of analyzing tabular data. □

# 6

## More on Cox Regression

Vital concepts like *time-dependent covariates, communal covariates*, handling of *ties, model checking, sensitivity analysis*, etc., are introduced in this chapter.

### 6.1 Time-Varying Covariates

Only a special kind of time-varying covariates can be treated in **R** by the packages eha and survival, and that is so-called *piecewise constant* functions. How this is done is best described by an example.

**Example 6.1** (Civil status)

The covariate (factor) civil status (called civst in the **R** data frame) is an explanatory variable in a mortality study, which changes value from 0 to 1 at marriage. How should this be coded in the data file? The solution is to create *two* records (for individuals that marry), each with a *fixed* value of *civ_st*:

1. Original record: $(t_0, t, d, x(s), t_0 < s \leq t)$, married at time $T$, $t_0 < T < t$:

$$\text{civst}(s) = \begin{cases} \text{unmarried}, & s < T \\ \text{married}, & s \geq T \end{cases}$$

2. First new record: $(t_0, T, 0, \text{unmarried})$, *always censored.*

3. Second new record: $(T, t, d, \text{married})$.

DOI: 10.1201/9780429503764-6

**TABLE 6.1** The coding of a time-varying covariate.

| id | enter | exit | event | civst |
|----|-------|------|-------|-----------|
| 23 | 0     | 30   | 0     | unmarried |
| 23 | 30    | 80   | 1     | married   |

The data file will contain two records like (with $T = 30$) what you can see in Table 6.1.

In this way, a time-varying covariate can always be handled by utilizing left truncation and right censoring. See also Figure 6.1 for an illustration.

**FIGURE 6.1** A time-varying covariate (unmarried or married). Right censored and left truncated at age 30 (at marriage).

And note that this situation is formally equivalent to a situation with *two individuals*, one unmarried and one married. The first one is right censored at exactly the same time as the second one enters the study (left truncation). □

A word of caution: Marriage status may be interpreted as an *internal* covariate, i.e., the change of marriage status is individual, and may depend on health status. For instance, maybe only healthy persons get married. So, the risk is that *health status* acts as a confounder in the relation between marriage and survival. Generally, the use of time dependent internal covariates is dangerous, and one should always think of possible confounding or reverse causality taking place when allowing for it.

## 6.2 Communal Covariates

Communal (external) covariates are covariates that vary in time outside any individual, and is common to all individuals at each specific *calendar* time. In econometric literature, such a variable is often called *exogenous*. This could be viewed upon as a special case of a time-varying covariate, but without the danger of reversed causality that was discussed above.

**Example 6.2** (Did food prices affect mortality in the 19th century?)

For some 19th century parishes in southern Sweden, yearly time series of food prices are available. In this example we use deviation from the log trend in annual rye prices, shown in Figure 6.2.

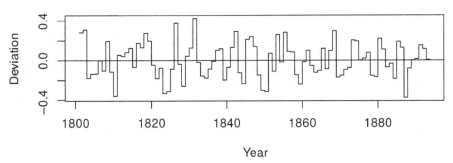

**FIGURE 6.2** Log rye price deviations from trend, 19th century southern Sweden.

The idea behind this choice of transformation is to focus on deviations from the "normal" as a proxy for *short-term variation in economic stress*. We illustrate the idea with a subset of the built-in data set scania, and we show the information for individual No. 1.

```
  id enter    exit event birthdate   sex    ses
29  1      50 59.242     1 1781.454 male lower
```

Now, to put on the food prices as a time-varying covariate, use the function `make.communal`. We show what happens to individual No. 1.

```
scand <- make.communal(scand, logrye[, 2, drop = FALSE],
                       start = 1801.75)
scand[scand$id == 1, ]
```

```
      enter    exit event birthdate id  sex   ses foodprices
1  50.000 50.296     0 1781.454  1 male lower      0.126
2  50.296 51.296     0 1781.454  1 male lower      0.423
3  51.296 52.296     0 1781.454  1 male lower     -0.019
4  52.296 53.296     0 1781.454  1 male lower     -0.156
5  53.296 54.296     0 1781.454  1 male lower     -0.177
6  54.296 55.296     0 1781.454  1 male lower     -0.085
7  55.296 56.296     0 1781.454  1 male lower     -0.007
8  56.296 57.296     0 1781.454  1 male lower      0.104
9  57.296 58.296     0 1781.454  1 male lower      0.118
10 58.296 59.242     1 1781.454  1 male lower     -0.197
```

A new variable, `foodprices` is added, and each individual's duration is split up in one year long intervals (except the first and last). This is the way of treating `foodprices` as a time-varying covariate. The analysis is then straightforward.

```
fit <- coxreg(Surv(enter, exit, event) ~ ses + sex + foodprices,
              data = scand)
```

Socio-economic status and sex do not matter much, but food prices do; the effect is almost significant at the 5 per cent level. See Table 6.2. □

**TABLE 6.2** Foodprices and old age mortality, 19th century Scania.

| Covariate | | Mean | Coef | Rel.Risk | S.E. | L-R p |
|---|---|---|---|---|---|---|
| ses | | | | | | 0.687 |
| | upper | 0.202 | 0 | 1 | (reference) | |
| | lower | 0.798 | −0.030 | 0.970 | 0.075 | |
| sex | | | | | | 0.500 |
| | male | 0.504 | 0 | 1 | (reference) | |
| | female | 0.496 | 0.041 | 1.042 | 0.061 | |
| foodprices | | 0.002 | 0.321 | 1.378 | 0.182 | 0.077 |
| Events | | 1086 | TTR | 26979 | | |
| Max. Log Likelihood | | −7184 | | | | |

## 6.3 Tied Event Times

Tied event times in theory cannot occur with continuous-time data, but it is impossible to measure duration and life lengths with infinite precision. Data are always more or less rounded, tied event times occur frequently in practice. This may cause problems (biased estimates) if occurring too frequently. There are a few ways to handle tied data, and the so-called exact method considers all possible permutations of the tied event times in each risk set. It works as shown in the following example.

**Example 6.3** (Likelihood contribution at tied event time)

$R_i = \{1, 2, 3\}$, 1 and 2 are events; two possible orderings:

$$
\begin{aligned}
L_i(\beta) &= \frac{\psi(1)}{\psi(1) + \psi(2) + \psi(3)} \times \frac{\psi(2)}{\psi(2) + \psi(3)} \\
&+ \frac{\psi(2)}{\psi(1) + \psi(2) + \psi(3)} \times \frac{\psi(1)}{\psi(1) + \psi(3)} \\
&= \frac{\psi(1)\psi(2)}{\psi(1) + \psi(2) + \psi(3)} \left\{ \frac{1}{\psi(2) + \psi(3)} + \frac{1}{\psi(1) + \psi(3)} \right\}
\end{aligned}
$$

☐

The main drawback with the exact method is that it easily becomes unacceptably slow, because of the huge number of permutations that may be necessary to consider. It is however available in the function `coxph` in the `survival` package as an option.

Fortunately, there are a few excellent approximations, most notably Efron's, which is the default method in most survival packages in **R**. Another common approximation, due to Breslow, is the default in some statistical software and also possible to choose in the `eha` and `survival` packages in **R**. Finally, there is no cost involved in using these approximations in the case of no ties at all; they will all give identical results.

With too much ties in the data, there is always the possibility to use discrete time methods. One simple way of doing it is to use the option `method = "ml"` in the `coxreg` function in the **R** package `eha`.

**Example 6.4** (Birth intervals.)

We look at length of birth intervals between first and second births for married women in 19th century northern Sweden, a subset of the data set 'fert', available in the `eha` package.
Four runs with `coxreg` are performed, with all the possible ways of treating ties.

```
library(eha)
first <- fert[fert$parity == 1, ]
## Default method (not necessary to specify method = "efron"
fit.e <- coxreg(Surv(next.ivl, event) ~ year + age, data = first,
            method = "efron")
## Breslow
fit.b <- coxreg(Surv(next.ivl, event) ~ year + age, data = first,
            method = "breslow")
## The hybrid mppl:
fit.mp <- coxreg(Surv(next.ivl, event) ~ year + age, data = first,
```

**TABLE 6.3** Comparison of four methods of handling ties, estimated coefficients.

|  | year | age |
|---|---|---|
| Efron | 0.0017 | −0.0390 |
| Breslow | 0.0017 | −0.0390 |
| MPPL | 0.0017 | −0.0391 |
| ML | 0.0017 | −0.0391 |

```
                 method = "mppl", coxph = FALSE)
## True discrete:
fit.ml <- coxreg(Surv(next.ivl, event) ~ year + age, data = first,
                 method = "ml", coxph = FALSE)
```

Then we compose a table of the four results, see Table 6.3.

There are almost no difference in results between the four methods. For the exact method mentioned above, the number of permutations is $\sim 7 \times 10^{192}$, which is impossible to handle. □

With the help of the function risksets in eha, it is easy to check the composition of risk sets in general and the frequency of ties in particular.

```
rs <- risksets(Surv(first$next.ivl, first$event))
tt <- table(rs$n.event)
tt
```

```
  1   2   3   4   5   6   7   9
369 199 133  60  27  14   2   2
```

This output says that 369 risk sets have only one event each (no ties), 199 risk sets have exactly two events each, etc.

The object `rs` is a `list` with seven components. Two of them are `n.event`, which counts the number of events in each risk set, and `size`, which gives the size (number of individuals under observation) of each risk set. Both these vectors are ordered after the `risktimes`, which is another component of the list. For the rest of the components, see the help page for `risksets`.

It is easy to produce the *Nelson-Aalen plot* with the output from `risksets`, see Figure 6.3.

```
par(las = 1)
plot(rs$risktimes, cumsum(rs$n.event / rs$size), type = "s",
     xlab = "Duration (years)", ylab = "Cum. hazards")
abline(h = 0)
```

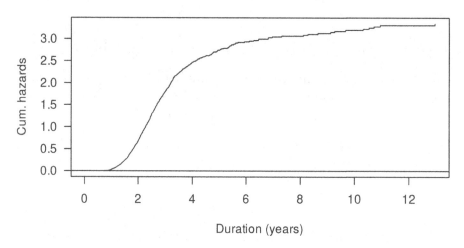

**FIGURE 6.3** Nelson-Aalen plot with risksets.

One version of the corresponding survival function is shown in Figure 6.4.

```
par(las = 1)
sur <- exp(-cumsum(rs$n.event / rs$size))
plot(rs$risktimes, sur, type = "s",
     xlab = "Duration (years)", ylab = "Surviving fraction")
abline(h = 0)
```

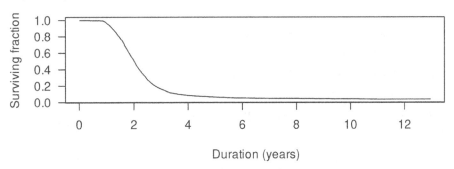

FIGURE 6.4 Survival plot with the aid of the function risksets.

## 6.4 Stratification

Stratification means that data is split up in groups called *strata*, and a separate partial likelihood function is created for each stratum, but with common values on the regression parameters corresponding to the common explanatory variables. In the estimation, these partial likelihoods are multiplied together, and the product is treated as a likelihood function. Thus, there is one restriction on the parameters, they are the same across strata.

There are typically two reasons for stratification. First, if the proportionality assumption does not hold for a factor covariate, a way out is to stratify along it. Second, a factor may have too many levels, so that it is inappropriate to treat is as an ordinary factor. This argument is similar to the one about using a frailty model (Chapter 10). Stratification is also a useful tool with matched data, see Chapter 11.

When a *factor* does not produce proportional hazards between categories, *stratify* on the categories. For tests of the proportionality assumption, see later in Section 6.7.1 of this chapter.

**Example 6.5** (Birth intervals)

For the birth interval data, we stratify on ses in the Cox regression:

```
library(eha)
source("R/fit.out.R")
fert1 <- fert[fert$parity == 1, ]
levels(fert1$parish) <- c("Jörn", "Norsjö", "Skellefteå")
fert1$parish <- relevel(fert1$parish, ref = "Skellefteå")
fit <- coxreg(Surv(next.ivl, event) ~ strata(ses) +
              I(age - 25) + I(year-1860) + prev.ivl + parish,
              data = fert1)
fit.out(fit,caption = "Birth intervals, 19th century Skellefteå.",
        label = "strafr6")
```

**TABLE 6.4** Birth intervals, 19th century Skellefteå.

| Covariate | Mean | Coef | Rel.Risk | S.E. | L-R p |
|-----------|------|------|----------|------|-------|
| I(age − 25) | 2.151 | −0.023 | 0.978 | 0.006 | NA |
| I(year − 1860) | −1.336 | −0.001 | 0.999 | 0.002 | 0.000 |
| prev.ivl | 1.309 | −0.225 | 0.799 | 0.029 | 0.750 |
| parish | | | | | 0.000 |
| Skellefteå | 0.937 | 0 | 1 | (reference) | |
| Jörn | 0.010 | 0.620 | 1.858 | 0.214 | |
| Norsjö | 0.053 | 0.187 | 1.206 | 0.109 | |
| Events | 1657 | TTR | 4500 | | |
| Max. Log Likelihood | −9047 | | | | |

The results in Table 6.4 show that calendar time is unimportant, while mother's age and the length of the previous interval, from marriage to first birth, both are important with birth intensity decreasing with both. Also, note that the continuous covariates are *centered* around a reasonable value in the range of both. It will not affect the rest of the table in this case, but with *interactions* involving either it would.

Next plot the fit, see Figure 6.5. The non-proportionality pattern is clearly visible. Note the line par(lwd = 1.5): It magnifies the *line widths* by 50 per cent in the plot, compared to default. And the values on the $y$ axis refers to a 25-year-old mother giving birth in Skellefteå 1860.

```
par(lwd = 1.5, cex = 0.8)
plot(fit, ylab = "Cumulative hazards", xlab = "Years")
```

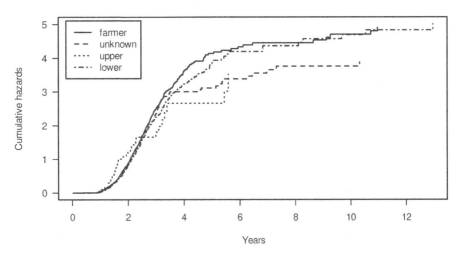

**FIGURE 6.5** Cumulative hazards by socio-economic status, second birth interval.

## 6.5  Sampling of Risk Sets

Some of the work by Langholz and Borgan (1995) is implemented in **eha**. The idea behind sampling of risk sets is that with huge data sets, in each risk sets there will be a few (often only one) events, but many survivors. From an efficiency point of view, not much will be lost by using only a random fraction of the survivors in each risk set. The gain is of course computational speed and memory saving. How it is done in **eha** is shown by an example.

**Example 6.6** (Sampling of risk sets, male mortality.)

For the male mortality data set, we compare a full analysis with one where only four survivors are sampled in each risk set.

```
fit <- coxreg(Surv(enter, exit, event) ~ ses,
              data = mort)
fit.4 <- coxreg(Surv(enter, exit, event) ~ ses,
              max.survs = 4, data = mort)
f1 <- coefficients(summary(fit))[c(1, 3)]
f4 <- coefficients(summary(fit.4))[c(1, 3)]
out <- rbind(f1, f4)
colnames(out) <- c("Coef", "se(Coef)")
rownames(out) <- c("Original", "Sample")
round(out, 4)
```

```
           Coef se(Coef)
Original -0.4795   0.1207
Sample   -0.5442   0.1357
```

The results are comparable and a gain in execution time was noted. Of course, with the small data sets we work with here, the difference is of no practical importance. □

It is worth noting that the function `risksets` has an argument `max.survs`, which, when it is set, sample survivors for each risk set in the data. The output can then be used as input to `coxreg`, see the relevant help pages for more information.

## 6.6 Residuals

Residuals are generally the key to judgment of model fit to data. It is easy to show how it works with *linear regression*.

**Example 6.7** (Linear regression)

Figure 6.6 shows a scatter plot of bivariate data and a fitted straight line (left panel) and residuals versus fitted values (right panel).

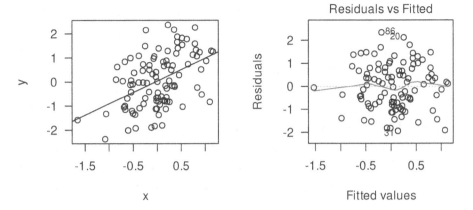

**FIGURE 6.6** Residuals in linear regression.

The residuals are the vertical distances between the line and the points. □

Unfortunately, this kind of simple and intuitive graph does not exist for results from a Cox regression analysis. However, there are a few ideas how to accomplish something similar. The main idea goes as follows. If $T$ is a survival time, and $S(t)$ the corresponding survivor function (continuous time), then it is a mathematical fact that $U = S(T)$ is uniformly distributed on the interval $(0, 1)$. Continuing, this implies that $-\log(U)$ is exponentially distributed with rate (parameter) 1.

It is thus shown that $H(T)$ ($H$ is the cumulative hazard function for $T$) is exponentially distributed with rate 1. This motivates the *Cox-Snell* residuals, given a sample $T_1, \ldots, T_n$ of survival times,

$$r_{Ci} = \hat{H}(T_i; \mathbf{x}_i) = \exp(\beta \mathbf{x}_i) \hat{H}_0(T_i), \quad i = 1, \ldots, n, \qquad (6.1)$$

which should behave like a censored sample from an exponential distribution with parameter 1, if the model is good. If, for some $i$, $T_i$ is censored, so is the residual.

### 6.6.1 Martingale residuals

A drawback with the Cox-Snell residuals is that they contain censored values. An attempt to correct the censored residuals leads to the so-called *martingale* residuals. The idea is simple; it builds on the "lack-of-memory" property of the exponential distribution. The expected value is 1, and by adding 1 to the *censored* residuals, they become predictions (estimates) of the corresponding uncensored values. then finally a twist is added: Subtract 1 from *all* values and multiply by –1, leading to the definition of the martingale residuals.

$$r_{Mi} = \delta_i - r_{Ci}, i = 1, \ldots, n. \tag{6.2}$$

where $\delta_i$ is the indicator of event for the $i$:th observation. These residuals may be interpreted as the *Observed* minus the *Expected* number of events for each individual. Direct plots of the martingale residuals tend to be less informative, especially with large data sets.

**Example 6.8** (Plot of martingale residuals)

The data set kidney from the **survival** package is used. For more information about it, read its help page in **R**. For each patient, two survival times are recorded, but only one will be used here. That is accomplished with the function duplicated.

```
library(survival)
head(kidney)
```

|   | id | time | status | age | sex | disease | frail |
|---|----|------|--------|-----|-----|---------|-------|
| 1 | 1  | 8    | 1      | 28  | 1   | Other   | 2.3   |
| 2 | 1  | 16   | 1      | 28  | 1   | Other   | 2.3   |
| 3 | 2  | 23   | 1      | 48  | 2   | GN      | 1.9   |
| 4 | 2  | 13   | 0      | 48  | 2   | GN      | 1.9   |
| 5 | 3  | 22   | 1      | 32  | 1   | Other   | 1.2   |
| 6 | 3  | 28   | 1      | 32  | 1   | Other   | 1.2   |

```
k1 <- kidney[!duplicated(kidney$id), ]
fit <- coxreg(Surv(time, status) ~ disease + age + sex,
              data = k1)
```

The *extractor function* residuals is used to extract the martingale
residuals from the fit. See Figure 6.7.

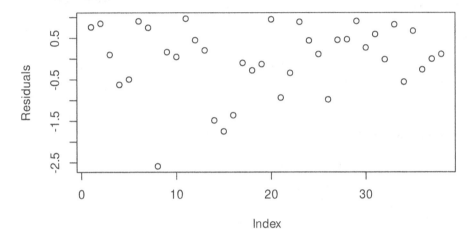

**FIGURE 6.7** Martingale residuals from a fit of the kidney data.

With a large data set, the plot will be hard to interpret, see Figure
6.8, which shows the martingale residuals from a fit of the *male
mortality* (mort) data in the package **eha**.

It is clear that the residuals are grouped in two clusters. The positive
ones are connected to the observations with an event, while the
negative ones are corresponding to the censored observations. It
is hard to draw any conclusions from plots like this one. However,
the residuals are used in functions that evaluate goodness-of-fit,
for instance the function cox.zph in the **survival** package. □

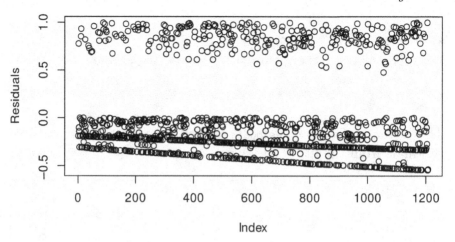

**FIGURE 6.8** Martingale residuals from a fit of the male mortality data.

## 6.7 Checking Model Assumptions

### 6.7.1 Proportionality

The proportionality assumption is extremely important to check in the proportional hazards models. We have already seen how violations of the assumption can be dealt with (stratification).

We exemplify with the 'births' data set in the **eha** package.

**Example 6.9** (Birth intervals, proportionality assumption)

In order to keep observations reasonably independent, we concentrate on one specific birth interval per woman, the interval between the second and third birth. That is, in our sample are all women with at least two births, and we monitor each woman from the day of her second delivery until the third, or if that never happens, throughout her observable fertile period, i.e, to age 50 or to loss of follow-up. The latter will result in a right-censored interval.

```
fert2 <- fert[fert$parity == 2, ]
```

Then we save the fit from a Cox regression performed by the function coxph in the **survival** package (important!),

```
fit <- survival::coxph(Surv(next.ivl, event) ~ ses + age +
                       year + parish,
          data = fert2)
```

and check the proportionality assumption. It is done with the function cox.zph in the **survival** package.

```
prop.full <- survival::cox.zph(fit)
prop.full
```

```
        chisq df      p
ses      9.76  3 0.0207
age      7.14  1 0.0075
year     2.34  1 0.1259
parish   1.39  2 0.4984
GLOBAL  18.73  7 0.0091
```

In the table above, look first at the last row, *GLOBAL*. It is the result of a test of the global null hypothesis that proportionality holds. The small $p$-value tells us that we have a big problem with the proportionality assumption. Looking further up, we see two possible problems, the variables age and ses. Unfortunately, age is a continuous variable. A categorical variable can be stratified upon, but now we have to categorize first.

Let us first investigate the effect of stratifying on ses.

```
fit1 <- coxph(Surv(next.ivl, event) ~ strata(ses) +
              age + year + parish, data = fert2)
```

and plot the result, see Figure 6.9.

The non-proportionality is visible, some curves cross. The test of proportionality of the stratified model shows that we still have a problem with age.

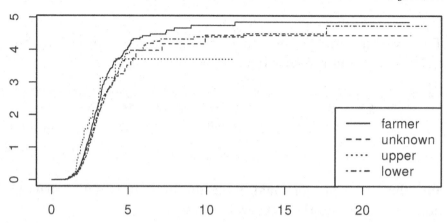

**FIGURE 6.9** Stratification on socio-economic status, third birth
interval data.

```
fit1.zph <- survival::cox.zph(fit1)
print(fit1.zph)
```

|        | chisq | df | p      |
|--------|-------|----|--------|
| age    | 7.85  | 1  | 0.0051 |
| year   | 1.61  | 1  | 0.2045 |
| parish | 1.47  | 2  | 0.4793 |
| GLOBAL | 9.19  | 4  | 0.0564 |

Since age is continuous covariate, we may have to categorize it.
First its distribution is checked with a histogram, see Figure 6.10.

```
hist(fert2$age, main = "", xlab = "age")
```

The range of age may reasonably be split into four equal-length
intervals with the cut function.

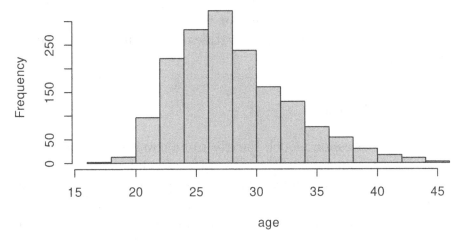

**FIGURE 6.10** The distribution of mother's age, third birth interval data.

```
fert2$qage <- cut(fert2$age, 4)
fit2 <- survival::coxph(Surv(next.ivl, event) ~ strata(ses) +
                    qage + year + parish, data = fert2)
```

and then the proportionality assumption is tested again.

```
fit2.zph <- survival::cox.zph(fit2)
fit2.zph
```

```
        chisq df     p
qage    7.589  3 0.055
year    0.758  1 0.384
parish  1.052  2 0.591
GLOBAL  8.536  6 0.201
```

The result is now reasonable, with the high age group deviating slightly. This is not so strange; fertility becomes essentially zero in the age range in the upper forties, and therefore it is natural that the proportionality assumption is violated. One way of dealing with the problem, apart from stratification, would be to analyze the age groups separately.  □

### 6.7.2  Log-linearity

In the Cox regression model, the effect of a covariate on the hazard is *log-linear*, that is, it affects the *log hazard* linearly. Sometimes a covariate needs to be transformed to fit into the pattern. The first question is then: Should the covariate be transformed in the analysis? A graphical test can be done by plotting the martingale residuals from a *null fit* against the covariate, See Figure 6.11.

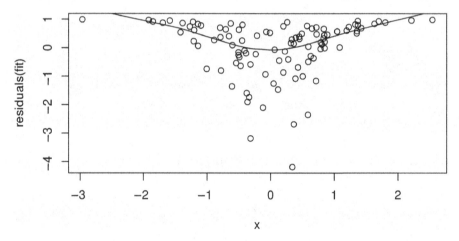

**FIGURE 6.11** Plot of residuals against the explanatory variable $x$.

The function `lowess` fits a "smooth" curve to a scatterplot. The curvature is clearly visible, let us see what happens if we make the plot with $x^2$ instead of $x$. (Note that we now are cheating!) See Figure 6.12.

```
plot(x^2, residuals(fit))
lines(lowess(x^2, residuals(fit)))
```

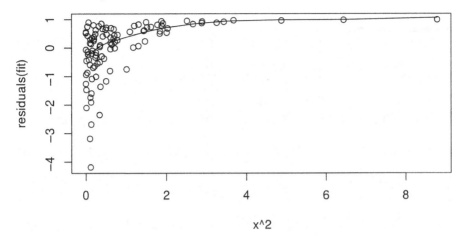

**FIGURE 6.12** Plot of residuals against the explanatory variable $x^2$.

This plot indicates more that an increasing $x$ is associated with an increasing risk for the event to occur.

## 6.8 Fixed Study Period Survival

Sometimes survival up to some fixed time is of interest. For instance, in medicine five-year survival may be used as a measure of the success of a treatment. The measure is then the probability for a patient to survive five years after treatment.

This is simple to implement; it is a binary experiment, where for each patient survival or not survival is recorded (plus covariates, such as treatment). The only complicating factor is incomplete observations, i.e., right censoring. In that case it is recommended to use ordinary Cox regression anyway. Otherwise, logistic regression with a suitable link function works well. Of course, as mentioned earlier, the *cloglog* link is what gets you closest to the proportional hazards model, but usually it makes very little difference. See Kalbfleisch and Prentice (2002), page 334 ff. for slightly more detail.

## 6.9   Left or Right Censored Data.

Sometimes the available data is either left or right censored, i.e., for each individual $i$, we know survival of $t_i$ or not, i.e., data

$$(t_i, \delta_i), \quad i = 1, \dots n,$$

$\delta_i = 1$ if survival (died after $t_i$) and $\delta_i = 0$ if not (died before $t_i$). It is still possible to estimate the survivor function $S$ nonparametrically. The likelihood function is

$$L(S) = \prod_{i=1}^{n} \{S(t_i)\}^{\delta_i} \{1 - S(t_i)\}^{1-\delta_i}$$

How to maximize this under the restrictions $t_i < t_j \Rightarrow S(t_i) \geq S(t_j)$ with the *EM algorithm* is shown by Groeneboom and Wellner (1992).

## 6.10   The Weird Bootstrap

In small to medium-sized samples, the estimated parameters and their standard errors may be not so reliable, and in such cases the bootstrap technique can be useful. There are a couple of ways to implement the technique in Cox regression, and here the *weird bootstrap* is presented.

The resampling procedure is done risk set by riskset. If a specific risk set of size $n$ contains $d$ events, $d$ "new" events are drawn from the $n$ members of the risk set, without replacement, and with probabilities proportional to their scores. Then the regression parameter is estimated and saved in the bootstrap sample. This procedure is repeated a suitable number of times.

The "weird" part of the resampling procedure is the fact that it is performed independently over risk sets, that is, one individual who is present in two risk sets may well have an event (for instance, die) in *both* risk sets. That is, the risk sets are treated as independent entities in this respect. See Andersen et al. (1993) for more detail.

The weird bootstrap is implemented in the `coxreg` function in the **eha** package. As an example, child mortality in 19th century Sweden is analyzed with respect to the sex difference in mortality.

```
fit <- coxreg(Surv(enter, exit, event) ~ sex, boot = 300,
              data = child)
b_hat <- fit$coefficents[1]
b_se <- sqrt(fit$var[1, 1])
b_sample <- fit$bootstrap[1, ]
```

This fitting procedure takes a fairly long time, around 19 minutes on a well equipped laptop (2018). The bootstrap result is a matrix where the number of rows is equal to the number of estimated parameters (one in this case), and the number of columns is equal to the number of bootstrap replicates (300 in this case). The `child` data set is fairly large with 26574 children and 5616 deaths. The number of risk sets is 2497, with sizes between 20000 and 27000, and up to 60 events in one risk set.

Now it is up to the user to do whatever she usually does with bootstrap samples. For instance, a start could be to get a grip of the distribution of the bootstrap sample around the parameter estimate, see Figure 6.13.

```
plot(density(b_sample - b_hat, bw = 0.005),
     xlab = "bootstrap - beta_hat", main = "")
abline(h = 0, v = 0)
```

The mean of the bootstrap sample is –0.0836 and its sample standard deviation is 0.0189. We can compare with the output of the standard Cox regression, see Table 6.5.

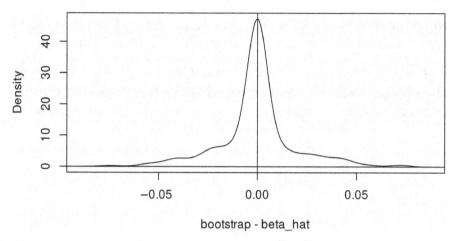

**FIGURE 6.13** Bootstrap sample distribution around the estimate.

**TABLE 6.5** Child mortality.

| Covariate | | Mean | Coef | Rel.Risk | S.E. | L-R p |
|-----------|---|------|------|----------|------|-------|
| sex | | | | | | 0.002 |
| | *male* | 0.510 | 0 | 1 | (reference) | |
| | *female* | 0.490 | −0.083 | 0.920 | 0.027 | |
| Events | | 5616 | TTR | 325030 | | |
| Max. Log Likelihood | | −56510 | | | | |

The mean is very close to the actual estimate, but the bootstrap standard error is slightly smaller than the estimated standard error.

For more on the bootstrap, see Efron and Tibshirani (1993), Hinkley (1988), and Hall and Wilson (1991).

# 7

# Register-Based Survival Data Models

During the last couple of decades, huge data bases with poulation data have popped up around the world. They are great sources of information in demographic and statistical research, but their sizes creates challenges for the statistical tools we traditionally use in the demographic analyses.

We distinguish between data sources with individual content and sources with tabular data. In both cases, however, we will end up analyzing tables, so we start by describing methods for analyzing tables.

Finally, in this chapter, we show how to combine the technique with communal covariates with tabulating data. The problem we solve is that the creation of an individual-based data set with communal covariates usually leads to uncomfortably large data sets. Tabulation is often a good solution to this problem.

## 7.1 Tabular Data

National statistical agencies produce statistics describing the population year by year, in terms of population size by age and sex, number of births by mother's age and sex of the newborn, number of deaths by age and sex, and so on. In the package **eha**, there are two data sets taken from Statistics Sweden[1], swepop and swedeaths.

The content of swepop is shown in Table 7.1 (first five rows out of 10504 rows in total).

---

[1]https://scb.se

DOI: 10.1201/9780429503764-7

**TABLE 7.1** First rows of 'swepop'.

| age | sex | year | pop |
|----:|-----|------|--------:|
| 0 | women | 1969 | 52673.0 |
| 0 | men | 1969 | 55728.0 |
| 1 | women | 1969 | 56831.0 |
| 1 | men | 1969 | 59924.0 |
| 2 | women | 1969 | 58994.5 |

**TABLE 7.2** First rows of 'swedeaths'.

| age | sex | year | deaths |
|----:|-----|------|-------:|
| 0 | women | 1969 | 491 |
| 0 | men | 1969 | 773 |
| 1 | women | 1969 | 33 |
| 1 | men | 1969 | 45 |
| 2 | women | 1969 | 22 |

It contains the average Swedish population size by year, age, and sex, where age is grouped into one-year classes, but the highest age class, labelled 100, is in fact "ages 100 and above". The original table gives the population sizes at the end of each year, but here we have calculated an average size over the year in question by taking the mean of the actual value and the corresponding value from the previous year. The population size can in this way be used as a proxy for, prediction of, "total time at risk" or *exposure*.

The content of swedeaths (first five rows) is shown in Table 7.2.

The (age, sex, year) columns in the two data frames are identical, so we can create a suitable data frame for analysis simply by putting swedeaths$deaths into the data frame swepop and rename it to sw. The first rows of sw are shown Table 7.3.

The response variable in our analysis is two-dimensional: (deaths, pop), which is on an "occurrence/exposure" form, suitable for an analysis with the function tpchreg. The result is a *proportional*

**TABLE 7.3** First rows of combined data frame.

| age | sex | year | pop | deaths |
|-----|-------|------|---------|--------|
| 0 | women | 1969 | 52673.0 | 491 |
| 0 | men | 1969 | 55728.0 | 773 |
| 1 | women | 1969 | 56831.0 | 33 |
| 1 | men | 1969 | 59924.0 | 45 |
| 2 | women | 1969 | 58994.5 | 22 |

*hazards* analysis with the assumption of a piecewise constant baseline hazard function:

```
system.time(fit <- tpchreg(oe(deaths, pop) ~ sex + I(year - 1995),
              time = age, last = 101, data = sw))
```

```
   user  system elapsed
  0.540   0.000   0.539
```

Note several things here:

1. The fitting procedure is wrapped in a call to the function `system.time`. This results in the first report headed "user system elapsed", giving the time in seconds for the whole operation ("elapsed"), split up by "user" (the time spent with our call) and "system" (time spent by the operating system, for instance reading, writing, and organizing the user operation). The interesting value for us is that the total elapsed time was around half a second. Quite OK for a proportional hazards regression analysis involving 4745063 events.

2. The response is `oe(deaths, pop)` where the function name `oe` is short for "occurrence/exposure".

3. All covariates have a *reference value*. For factors with the default coding, the reference category is given, but for

continuous covariates we need to specify it, if we are not satisfied with the default value *zero*. In the case with the covariate year, we are certainly *not* satisfied with zero as reference point, so we choose a suitable value in the range of the observed values. This is done in practice by *subtracting* the reference value from the corresponding variable, forming year - 1995 in our case. In the output below, this will only affect the value of *Restricted mean survival*, which is calculated for a "reference individual", in this case *a woman living her life during the year 1995*. There are not many such women, of course, what we are doing here is comparing periods, not cohorts.

4.  Note the argument last = 101. Since we are fitting a survival distribution that is unbounded in time, we should specify the range for our data. The lower limit is given by min(age) (the argument time), but there is no natural upper limit. The choice of the value for last do not affect parameter estimates and general conclusions, it only affects plots in the extreme of the right tail (slightly).

**TABLE 7.4** Swedish mortality, 1968–2020.

| Covariate | | Mean | Coef | Rel.Risk | S.E. | L-R p |
|---|---|---|---|---|---|---|
| sex | | | | | | 0.000 |
| | *women* | 0.503 | 0 | 1 | (reference) | |
| | *men* | 0.497 | 0.440 | 1.553 | 0.001 | |
| I(year − 1995) | | 0.520 | −0.015 | 0.985 | 0.000 | 0.000 |
| Events | | 4745063 | TTR | 4.6e+08 | | |
| Max. Log Likelihood | | −19283067 | | | | |
| Restricted mean survival: | | 80.9 | in (0, 101] | | | |

Note the "Restricted mean survival": It gives the expected remaining life *at birth* for a *woman* in the *year 1995* (the reference values).

The graph of the baseline hazard function ("age-specific mortality") is shown in Figure 7.1.

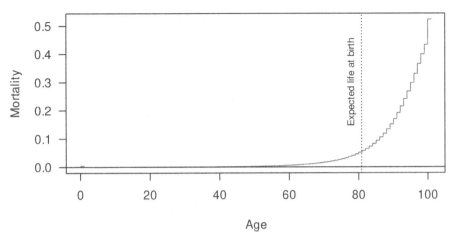

**FIGURE 7.1** Baseline hazard function, women 1995. Model based.

Apart from the first year of life, mortality seems to be practically zero the first forty years of life, and then exponentially increasing with age.

This is simple enough, but some questions remain unanswered. First, the model assumes proportional hazards, implying for instance that male mortality is around 50 per cent higher than female in all ages. It is fairly simple to organize a formal test of the proportionality assumption, but it is close to meaningless, because of the huge amount of data (almost five million deaths, "all" null hypotheses will be rejected). A better alternative is to do it graphically by *stratifying* on sex.

Due to the extremely low mortality in early ages, Figure 7.2 is quite non-informative. We try the same plot on a log scale, see Figure 7.3.

The female advantage is largest in early adulthood, after which it slowly decreases. An even clearer picture is shown in Figure 7.4, where the ratio of male vs. female mortality is plotted.

This graph gives the clearest picture of the relative differences in mortality over the life span. We see, somewhat unexpectedly, that the drop in the ratio is broken around age 50, and then continues after age 70. This is not clearly visible in the earlier figures. A possible explanation is that the age interval 20–50 approximately

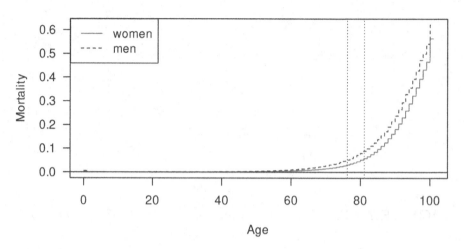

**FIGURE 7.2** Baseline hazard function, women and men 1995.

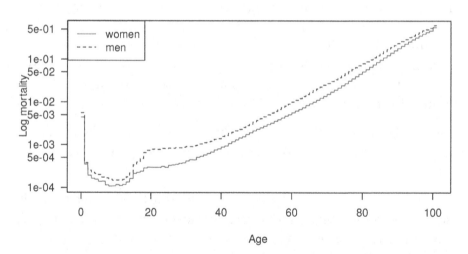

**FIGURE 7.3** Baseline hazard functions, women and men 1995. Log scale.

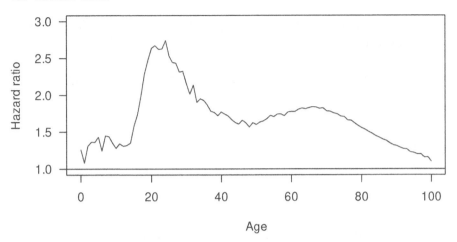

**FIGURE 7.4** Hazard ratio by age, men vs. women 1995.

coincides with women's fertility period, that is, during this period, the female mortality advantage decreases more than pure age motivates.

This pattern may change over the time period 1969–2020. We split the data by year into quarters in order to check this.

The rise after age 50 seems to vanish with time, according to Figure 7.5.

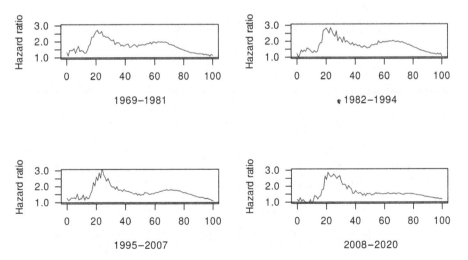

**FIGURE 7.5** Baseline hazard ratios, men vs. women by time period.

Finally, we investigate the development of expected life ($e_0$) over time, see Figure 7.6.

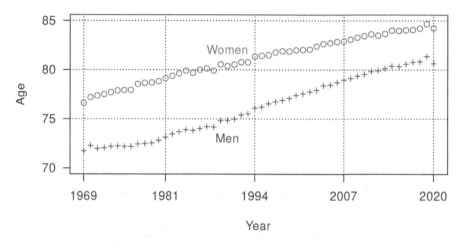

**FIGURE 7.6** Expected life at birth by sex and year (period data).

The difference in longevity between women and men is slowly decreasing over the time period. Also note the values in the last two years: Mortality in Sweden was unexpectedly low in the year 2019, resulting in high values of expected life, but in 2020, the pandemic virus *covid-19* struck, and the expected life dropped by around five months for women and almost nine months for men, compared to 2019.

## 7.2   Individual Data

The analysis of individual survival data is what the rest of this book is all about, so what is new here? The answer lies in the *size* of the data sets, that is, the number of individuals is *huge*.

As an example, we have mortality data consisting of all Swedish citizens aged 30 and above for the years 2001–2013. It may be

**TABLE 7.5** Register data, Sweden December 31, 2001.

| id | year | sex | birthdate | deathdate | civst | ses | region |
|----|------|-----|-----------|-----------|-------|-----|--------|
| 3 | 2001 | female | 1967-02-15 | NA | unmarried | worker | 21 |
| 5 | 2001 | female | 1928-11-16 | NA | prev.married | middle | 19 |
| 6 | 2001 | female | 1919-02-16 | 2013-01-27 | prev.married | lower | 1 |
| 9 | 2001 | female | 1935-11-16 | NA | married | elite | 24 |

described as yearly censuses, where all Swedish persons alive at December 31 the specific year are registered together with vital information. It starts with register data of the form shown in Table 7.5.

The variables are

- **id** A unique id number, which allows us to link information from different sources (and years) to one individual.
- **year** The present year, 2001, ..., 2013.
- **sex** Female or male.
- **birthdate** A constructed variable. Due to integrity reasons, birth time information is only given in the form of year and quarter of the year. So the *birthdate* is created by year and the first quarter is represented by the decimal places ".125" (equal roughly to February 15), the second quarter is ".375", third ".625", and finally the fourth quarter is represented by ".875"). The birthdate will be off by plus or minus one month and a half, which is deemed to be a good enough precision concerning our analysis of adult mortality.
- **deathdate** Is given by day in the sources. The missing value symbol (*NA*) is given to those who are not registered as dead before January 1, 2014. Each date is converted to decimal form.
- **civst, ses, region** Information about *civil status*, *socioeconomic status*, and *geographic area*, respectively.

From this yearly information, survival data is created by following each individual one year ahead, until the next "census". Age at

**TABLE 7.6** Constructed data, Sweden December 31, 2001.

| id | year | sex | birthdate | deathdate | enter | exit | event |
|---|---|---|---|---|---|---|---|
| 3 | 2001 | female | 1967.125 | NA | 34.875 | 35.875 | 0 |
| 5 | 2001 | female | 1928.875 | NA | 73.125 | 74.125 | 0 |
| 6 | 2001 | female | 1919.125 | 2013.074 | 82.875 | 83.875 | 0 |
| 9 | 2001 | female | 1935.875 | NA | 66.125 | 67.125 | 0 |

start, enter, is calculated by subtracting birthdate from the date of presence, year + 1. For an individual with a deathdate less than or equal to enter + 1 = year + 2, the variable exit is calculated as deathdate - birthdate and the variable event is set equal to one, otherwise exit = enter + 1 and event = 0, a censored observation.

The result of this is shown for the same individuals as above in Table 7.6, with three covariates omitted in order to save space.

This file from the year 2001 (covering the year 2002!) contains *5.8 million* individuals and 92 thousand deaths. To perform a Cox regression on such massive data sets (and there are thirteen of them) is time, software, and hardware consuming, but there is a simple remedy: Make tables and fit *piecewise constant proportional hazards* (pch) models.

This is done with the aid of the eha function toTpch. We decide to choose the pch model as constant over five-year intervals 30–35, 35–40, ..., 100–105, and we replace birthdate by birth year (more suitable as a covariate) and call it cohort. It is done with the aid of the R function floor:

```
lisa2001$cohort <- floor(lisa2001$birthdate)
```

The function floor simply removes the decimals from a positive number. Then we apply the function toTpch:

```
listab2001 <- toTpch(Surv(enter, exit, event) ~ sex + cohort + year +
                                         civst + region + ses,
                     data = lisa2001,
                     cuts = c(seq(30, 100, by = 5), 105))
```

**TABLE 7.7** Mortality in ages above 30, Sweden 2002. 'year' = 2001).

| sex | cohort | year | civst | ses | age | event | exposure | region |
|-----|--------|------|-------|-----|-----|-------|----------|--------|
| male | 1967 | 2001 | unmarried | elite | 30−35 | 0 | 869.375 | 1 |
| female | 1967 | 2001 | unmarried | elite | 30−35 | 0 | 538.875 | 1 |
| male | 1968 | 2001 | unmarried | elite | 30−35 | 0 | 1923.000 | 1 |
| female | 1968 | 2001 | unmarried | elite | 30−35 | 0 | 1286.000 | 1 |
| male | 1969 | 2001 | unmarried | elite | 30−35 | 0 | 2045.000 | 1 |

The result is shown in Table 7.7 (four first rows) which is a table with 35998 rows. That is, one row for each unique combination of the six covariates for each interval, empty combinations excluded. Note the three created variables, event which is the total number of events for each combination of covariates, exposure, which is the total time at risk for each corresponding combination, and age, which is the age intervals with constant hazard.

In the survival analysis, the pair (event, exposure) is the *response*, and behind the scene, *Poisson regression* is performed with event as response and log(exposure) as *offset*.

Finally, thirteen similar tables are created (one for each year 2001–2013) and merged, using the **R** command rbind.

A summary of the result gives

- Number of records: 489783.
- Number of deaths: 1.16 millions.
- Total time at risk: 76.75 million years.

Now this data set is conveniently analyzed with the aid of the function tpchreg in eha. An example:

```
listab$year <- listab$year - 2000 # Center!
list2 <- aggregate(cbind(event, exposure) ~ sex + civst + ses +
                    year + age, FUN = sum, data = listab)
```

```
fit <- tpchreg(oe(event, exposure) ~ strata(sex) + civst + ses +
                year, data = list2, time = age, last = 105)
xx <- summary(fit)
emen <- round(xx$rmean[1], 2)
ewomen <- round(xx$rmean[2], 2)
```

**TABLE 7.8** Mortality in Sweden 2002–2014, ages 30 and above.

| Covariate | | **Mean** | **Coef** | **Rel.Risk** | **S.E.** | **L-R p** |
|---|---|---|---|---|---|---|
| civst | | | | | | 0.000 |
| | *married* | 0.512 | 0 | 1 | (reference) | |
| | *unmarried* | 0.264 | 0.482 | 1.619 | 0.003 | |
| | *prev.married* | 0.224 | 0.327 | 1.386 | 0.002 | |
| ses | | | | | | 0.000 |
| | *worker* | 0.340 | 0 | 1 | (reference) | |
| | *elite* | 0.115 | −0.327 | 0.721 | 0.006 | |
| | *middle* | 0.318 | −0.168 | 0.845 | 0.004 | |
| | *lower* | 0.228 | 0.363 | 1.437 | 0.003 | |
| year | | 7.088 | −0.002 | 0.998 | 0.000 | 0.000 |
| Events | | 1161164 | TTR | 76752806 | | |
| Max. Log Likelihood | | −4538560 | | | | |
| Restricted mean survival in | (30, 105] : | | | | | |
| | | male | female | | | |
| | | 52.57 | 57.02 | | | |

The result is shown in Table 7.8.

We note a couple of things here:

1. The data frame listab contains 489783 records, and the execution time of a Poisson regression might take a long time. Since we did not use the covariates region (a factor with 25 levels) and cohort (92 levels), we could aggregate over these variables which resulted in the data frame list2 with only 4493 records. This cuts down computing time by approximately 99 per cent, or to 0.17 seconds from 17 seconds. The results are not affected by this. Also note

the similarity between the `formula` arguments of the two functions `aggregate` and `tpchreg`. The differences are that in `aggregate`, the function `cbind` is used instead of `oe`, and `strata()` is left out.

2. All *p*-values are effectively zero, a consequence of of the huge amount of data. It is also reasonable to claim that this constitutes a *total investigation*, and therefore *p*-values are meaningless.

3. The restricted mean survival *at age 30* for a *married worker* in the *year 2002* is 52.57 years for men and 57.02 years for women, that is, the expected age of death is 82.57 years for men and 87.02 years for women alive at 30.

A graph of the conditional survival functions is shown in Figure 7.7.

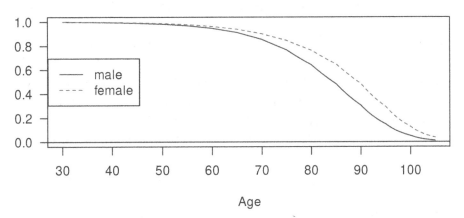

**FIGURE 7.7** Survival above 30, males and females, married worker, Sweden 2002.

## 7.3 Communal Covariates and Tabulation

The application of the function make.communal (see previous chapter) will often result in impractically large data sets, and it is recommended to combine it with the tabulation technique suggested in this chapter. We illustrate it with an investigation of the influence of weather conditions on mortality for two time periods, 1901–1950 and 1990–2014. In the first case focus is on the effect of low temperature on mortality, and the second case has the opposite focus, is an extremely hot summer bad for survival. The geographical area under study is the Swedish town Umeå, situated at the Gulf of Bothnia on the latitude 64 degrees north.

### 7.3.1 Temperature and mortality, Umeå 1901–1950

Daily temperature data from SMHI[2] for central Umeå, starting at November 1, 1858 and ending at September 30, 1979 (this particular weather station was replaced at this last date) are used together with population data from Umeå. Since we attempt to match these data to population data for the time period January 1, 1901–December 31, 1950, we cut the temperature data to the same time period (see Table 7.9).

```
library(eha)
temp <- read.table("~/Forskning/Data/ume_temp.csv",
                   header = TRUE,
                   skip = 10, sep = ",")[, 1:3]
## We need only the first three columns
names(temp) <- c("date", "time", "temp")

temp$date <- as.Date(temp$date)
```

[2]https://www.smhi.se/data/meteorologi/ladda-ner-meteorologiska-observationer#param=airtemperatureInstant,stations=all,stationid=140500

**TABLE 7.9** Temperature in Umeå, 1901−1950.

| date | time | temp | year | month | quarter |
|------|------|------|------|-------|---------|
| 1901-01-01 | 07:00:00 | −17.5 | 1901 | 1 | Q1 |
| 1901-01-01 | 13:00:00 | −17.6 | 1901 | 1 | Q1 |
| 1901-01-01 | 20:00:00 | −22.1 | 1901 | 1 | Q1 |
| 1901-01-02 | 07:00:00 | −13.1 | 1901 | 1 | Q1 |
| 1901-01-02 | 13:00:00 | −12.2 | 1901 | 1 | Q1 |
| 1901-01-02 | 20:00:00 | −18.9 | 1901 | 1 | Q1 |
| 1901-01-03 | 07:00:00 | −18.4 | 1901 | 1 | Q1 |
| 1901-01-03 | 13:00:00 | −17.1 | 1901 | 1 | Q1 |
| 1901-01-03 | 20:00:00 | −18.3 | 1901 | 1 | Q1 |
| 1901-01-04 | 07:00:00 | −16.3 | 1901 | 1 | Q1 |

```
temp$year <- as.numeric(format(temp$date, "%Y"))
temp$month <- as.numeric(format(temp$date, "%m"))
temp <- temp[order(temp$date, temp$time), ]
temp$quarter <- quarters(temp$date)
temp <- temp[temp$year %in% 1901:1950, ]
##

temp$time <- as.factor(temp$time)
temp$quarter <- as.factor(temp$quarter)
```

Apparently, there are three measurements each day, and the temperature range is −34.8, 30 degrees Celsius. The variables *year*, *month*, and *quarter* were extracted from the *date*.

We extract population data for Umeå, Jan 1 1901 to Dec 31 1950, and include all ages between 15 and 100, see Table 7.10 for the first few rows.

There are 186033 records describing 73780 individuals in this selection. In order to simplify the presentation, only two covariates, sex and urban, are included.

The first question is what to do with the temperature measurements. They span 50 years, with mostly three measurements per day. In this

**TABLE 7.10** Umeå data, ages 15−100, time period 1901−1950.

| id | sex | birthdate | enter | exit | event | socStatus |
|----|-----|-----------|-------|------|-------|-----------|
| 19 | male | 1922-09-28 | 26.489 | 27.166 | 0 | high |
| 31 | female | 1888-10-08 | 18.448 | 18.453 | 0 | low |
| 31 | female | 1888-10-08 | 18.453 | 20.129 | 0 | low |

**TABLE 7.11** Maximum daily temperature, Umeå 1901−1950.

| date | maxtemp | month | quarter | year |
|------|---------|-------|---------|------|
| 1901-01-01 | −17.5 | 1 | Q1 | 1901 |
| 1901-01-02 | −12.2 | 1 | Q1 | 1901 |
| 1901-01-03 | −17.1 | 1 | Q1 | 1901 |

first analysis we focus on the effect on mortality of *low* temperatures, so we start by taking the maximum of daily temperatures, then minimum over month. We thus end up with one measurement per month and year, in total 600 measurements.

Start by finding the *maximum* value each day, result in Table 7.11.

```
idx <- with(temp, tapply(date, date))
maxtemp <- with(temp, tapply(temp, date, max))
temp$maxtemp <- maxtemp[idx]
## Reduce to one measurement per day:
mtemp <- temp[!duplicated(temp$date),
            c("date", "maxtemp", "month", "quarter", "year")]
rownames(mtemp) <- 1:NROW(mtemp)
tbl(head(mtemp, 3), fs = 11,
    caption = "Maximum daily temperature, Umeå 1901-1950.")
```

The next step is to take *minimum* over year and month, see Table 7.12.

**TABLE 7.12** Average daily max temperature by month, Umeå 1901–1950.

| month | year | temp |
|------:|-----:|-----:|
| 1 | 1901 | −17.5 |
| 2 | 1901 | −15.1 |
| 3 | 1901 | −12.2 |

```
atmp <- aggregate(list(temp = mtemp$maxtemp),
                  by = mtemp[, c("month", "year")],
                  FUN = min)
tbl(head(atmp, 3), fs = 10,
    caption = "Average daily max temperature by month, Umeå 1901-50.")
```

Now we can apply `make.communal` to our data and split spells *by year and month*. The result is shown in Table 7.13.

```
atmp$yearmonth <- atmp$year * 100 + atmp$month
## A trick to get format 190101
comtime <- system.time(
    nt <- make.communal(ume, atmp["yearmonth"],
                        start = 1901, period = 1/12))

nt$enter <- round(nt$enter, 3)
nt$exit <- round(nt$exit, 3)
oj <- nt$enter >= nt$exit - 0.0001 # Too short interval!
nt$enter[oj] <- nt$exit[oj] - 0.001  # Break accidental ties
nt$year <- nt$yearmonth %/% 100 # Restore 'year'.
nt$month <- nt$yearmonth %% 100 # Restore 'month'.
saveRDS(nt, file = "mytables/nt.rds")
nt <- nt[nt$month %in% c(12, 1, 2), ]
tbl(head(nt[, -8], 3), fs = 10,
    caption = "Population data by year and month, Umeå 1901-50.")
```

Total time for *making communal*: 25 seconds.

**TABLE 7.13** Population data by year and month, Umeå 1901–50.

| enter | exit | event | birthdate | id | sex | socStatus | year | month |
|---|---|---|---|---|---|---|---|---|
| 19.145 | 19.228 | 0 | 1888.772 | 31 | female | low | 1907 | 12 |
| 19.228 | 19.312 | 0 | 1888.772 | 31 | female | low | 1908 | 1 |
| 19.312 | 19.395 | 0 | 1888.772 | 31 | female | low | 1908 | 2 |

**TABLE 7.14** Temperature added to Umeå data.

| id | sex | enter | exit | event | year | month | socStatus | temp |
|---|---|---|---|---|---|---|---|---|
| 31 | female | 19.145 | 19.228 | 0 | 1907 | 12 | low | -18.2 |
| 31 | female | 19.228 | 19.312 | 0 | 1908 | 1 | low | -23.2 |
| 31 | female | 19.312 | 19.395 | 0 | 1908 | 2 | low | -10.8 |

The next step is to read temperature from atmp to nt, see Table 7.14 (selected variables).

```
indx <- match(nt$yearmonth, atmp$yearmonth)
nt$temp <- atmp$temp[indx]
```

This data frame contains 2708779 records, maybe too much for a comfortable Cox regression. Let's see what happens (Table 7.15):

```
ptm <- proc.time()
fit <- coxreg(Surv(enter, exit, event) ~
              sex + socStatus + temp, data = nt)
ctime <- proc.time() - ptm
fit.out(fit, label = "coxph7",
        caption = "Mortality and temperature, Umeå 1901-50.")
```

This is quite bad (the computing time, almost 2 minutes), but we already have a substantial result: Survival chances *increase* with temperature. But remember that Umeå is a northern town, not

**TABLE 7.15** Mortality and temperature, Umeå 1901–50.

| Covariate | | Mean | Coef | Rel.Risk | S.E. | L-R p |
|---|---|---|---|---|---|---|
| sex | | | | | | 0.301 |
| | *male* | 0.490 | 0 | 1 | (reference) | |
| | *female* | 0.510 | −0.038 | 0.963 | 0.037 | |
| socStatus | | | | | | 0.331 |
| | *low* | 0.501 | 0 | 1 | (reference) | |
| | *high* | 0.499 | 0.036 | 1.037 | 0.037 | |
| temp | | −15.430 | −0.006 | 0.994 | 0.003 | 0.081 |
| Events | | 3022 | TTR | 221241 | | |
| Max. Log Likelihood | | −21893 | | | | |

much warm weather here. And further, no signs of interactions with sex or social status (not shown).

The analyses so far are very rudimentary, including all ages in one and so on. Here we show how extract small subsets of the full data set by some simple and reasonable assumptions. It builds on *categorical covariates* and an assumption of *piecewise constant hazards*. We use the function toTpch in the eha package, see Table 7.16.

```
told <- toTpch(Surv(enter, exit, event) ~ sex + socStatus +
                yearmonth,
             data = nt, cuts = seq(15, 100, by = 5))
indx <- match(told$yearmonth, atmp$yearmonth)
told$temp <- atmp$temp[indx]
saveRDS(told, file = "mytables/told73.rds")
##told <- readRDS("mytables/told73.rds")
tbl(head(told, 6), fs = 11,
    caption = "Tabular Umeå 1901-1950 data.")
```

The tabulation reduces the data set from 2.5 million records to 19 thousands. And computing time accordingly (reduced to a quarter of a second!), without jeopardizing the results, see Table 7.17.

**TABLE 7.16** Tabular Umeå 1901−1950 data.

| sex | socStatus | yearmonth | age | event | exposure | temp |
|-----|-----------|-----------|-----|-------|----------|------|
| male | low | 190101 | 15−20 | 0 | 37.393 | −17.5 |
| female | low | 190101 | 15−20 | 2 | 37.869 | −17.5 |
| male | high | 190101 | 15−20 | 1 | 37.408 | −17.5 |
| female | high | 190101 | 15−20 | 0 | 36.573 | −17.5 |
| male | low | 190102 | 15−20 | 1 | 37.221 | −15.1 |
| female | low | 190102 | 15−20 | 0 | 37.589 | −15.1 |

**TABLE 7.17** Temperature and mortality, Umeå 1901−1950. PCH model.

| Covariate | | Mean | Coef | Rel.Risk | S.E. | L-R p |
|-----------|------|------|------|----------|------|-------|
| sex | | | | | | 0.318 |
| | *male* | 0.490 | 0 | 1 | (reference) | |
| | *female* | 0.510 | −0.037 | 0.964 | 0.037 | |
| socStatus | | | | | | 0.347 |
| | *low* | 0.501 | 0 | 1 | (reference) | |
| | *high* | 0.499 | 0.035 | 1.035 | 0.037 | |
| temp | | −15.430 | −0.006 | 0.994 | 0.003 | 0.086 |
| Events | | 3022 | TTR | 221238 | | |
| Max. Log Likelihood | | −13591 | | | | |
| Restricted mean survival: | | 51.9 | in (15, 100] | | | |

Essentially the same conclusion as in the earlier attempts: Mortality decreases with increasing temperature.

### 7.3.2   Hot summers in Umeå, 1990–2014

As the next example of combining the use of the function `eha::make.communal` and tabulation, the effect of hot summers on mortality is examined. Data at hand relate to Umeå during the time period 1990–2014.

**TABLE 7.18** First rows of temperature data for Umeå 1990−2014.

| day | hour | temp | year | month |
|-----|------|------|------|-------|
| 1990-01-01 | 0 | −4.5 | 1990 | 1 |
| 1990-01-01 | 1 | −5.2 | 1990 | 1 |
| 1990-01-01 | 2 | −7.1 | 1990 | 1 |
| 1990-01-01 | 3 | −9.2 | 1990 | 1 |
| 1990-01-01 | 4 | −9.6 | 1990 | 1 |

**TABLE 7.19** First rows of individuals from Umeå 1990−2014.

| birthdate | sex | enter | exit | event | year |
|-----------|-----|-------|------|-------|------|
| 1935.875 | female | 54.125 | 79.124 | 0 | 1987 |
| 1977.125 | female | 30.872 | 33.874 | 0 | 2008 |
| 1969.125 | female | 26.872 | 30.873 | 0 | 1996 |
| 1937.875 | female | 52.125 | 77.124 | 0 | 1987 |
| 1980.125 | female | 23.872 | 25.874 | 0 | 2004 |
| 1911.125 | female | 78.875 | 87.569 | 1 | 1987 |

Temperature data for different places in Sweden can be downloaded from SMHI − The Swedish Meteorological and Hydrological Institute[3], data for Umeå is read and presented in Table 7.18.

There are 24 measurements taken every day, with few exceptions, in total 184398 values.

Population data for Umeå are extracted from the Longitudinal integrated database for health insurance and labour market studies (LISA)[4]. LISA contains information about all Swedes aged 16 and above on 31 December yearly, starting at 1990. See Table 7.19.

Then perform the aggregation, see Table 7.20.

The next step is to aggregate the population data, but we must first split it by month and year, using the function make.communal. We want to keep track of year and month, so the simplest way

---

[3]https://www.smhi.se

[4]https://www.scb.se

**TABLE 7.20** Aggregated summer temperature data, Umeå 1990—2014.

| month | year | temp | yearmonth |
|-------|------|------|-----------|
| June | 1990 | 23.3 | 199006 |
| July | 1990 | 24.8 | 199007 |
| August | 1990 | 23.6 | 199008 |
| June | 1991 | 19.3 | 199106 |
| July | 1991 | 28.6 | 199107 |
| August | 1991 | 25.5 | 199108 |

to do it is to use the variable yearmonth in temp as our communal covariate. And then read temperature from temp:

```
compop <- eha::make.communal(pop, aggrtemp["yearmonth"],
                    start = 1990, period = 1 / 12)
indx <- match(compop$yearmonth, aggrtemp$yearmonth)
compop$temp <- aggrtemp$temp[indx]
compop$year <- compop$yearmonth %/% 100
compop$month <- compop$yearmonth %% 100
```

```
out <- toTpch(Surv(enter, exit, event) ~ yearmonth + sex,
         data = compop,
              cuts = c(seq(15, 95, by = 5), 110))
indx <- match(out$yearmonth, aggrtemp$yearmonth)
out$temp <- aggrtemp$temp[indx]
```

Then the analysis of female mortality in the ages 80 and above and its dependence on high temperature (above 29 degrees Celsius), see Table 7.21. The July maximum temperatures by year are shown in Figure 7.8.

With individual-based data and Cox regression, see Table 7.22.

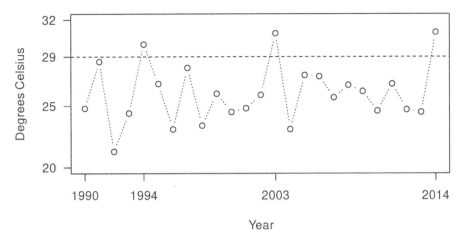

**FIGURE 7.8** Maximum temperatures in July by year 1990–2014.

**TABLE 7.21** Tabular PH analysis, female (age 80+) mortality and high temperature, Umeå 1990–2014.

| Covariate | | Mean | Coef | Rel.Risk | S.E. | L-R p |
|---|---|---|---|---|---|---|
| I(temp > 29) | | | | | | 0.014 |
| | FALSE | 0.876 | 0 | 1 | (reference) | |
| | TRUE | 0.124 | 0.326 | 1.386 | 0.128 | |
| Events | | 440 | TTR | 4580 | | |
| Max. Log Likelihood | | −1413 | | | | |
| Restricted mean survival: | | 9.47 | in (80, 110] | | | |

The results in the two tables are almost identical, women above 80 years of age are vulnerable to extreme temperatures: The risk increases with around 30 per cent (95 per cent confidence interval: from 7 to 77 per cent, quite wide, but a *p*-value of 1 per cent).

**TABLE 7.22** Cox regression, female (age 80+ ) mortality and high temperature, Umeå 1990—2014.

| Covariate | Mean | Coef | Rel.Risk | S.E. | L-R p |
|---|---|---|---|---|---|
| I(temp > 29) | | | | | 0.017 |
| *FALSE* | 0.876 | 0 | 1 | (reference) | |
| *TRUE* | 0.124 | 0.319 | 1.376 | 0.128 | |
| Events | 440 | TTR | 4580 | | |
| Max. Log Likelihood | −2863 | | | | |

# 8

## Parametric Models

We have already been given an example of a parametric survival model, the *piecewise constant hazards model* in the previous chapter. Apart from that, we have so far studied nonparametric methods for survival analysis. This is a tradition that has its roots in medical (cancer) research. In technical applications, on the other hand, parametric models dominate; of special interest is the *Weibull* model (Weibull, 1951). The text book by Lawless (2003) is a good source for the study of parametric survival models. More technical detail about parametric distributions is found in Appendix B .

Three kinds of parametric models are considered here: the *proportional hazards* and the *accelerated failure time* in continuous time, and the *discrete time proportional hazards* models.

## 8.1 Proportional Hazards Models

A proportional hazards family of distributions is generated from one specific continuous distribution by multiplying the hazard function of that distribution by a strictly positive constant, and letting that constant vary over the full positive real line. So, if $h_0$ is the hazard function corresponding to the generating distribution, the family of distributions can be described by saying that $h$ is a member of the family if

$$h_1(t) = ch_0(t) \quad \text{for some } c > 0, \text{ and all } t > 0.$$

DOI: 10.1201/9780429503764-8

Note that it is possible to choose any hazard function (on the positive real line) as the generating function. The resulting proportional hazards class of distributions may or may not be a well recognized family of distributions.

In **R**, **eha** is one of the packages that can fit parametric proportional hazards models. In the following subsections, the possibilities are examined.

The parametric distribution functions that naturally can be used as the baseline distribution in the function phreg are the *Weibull*, *Extreme value*, and the *Gompertz* distributions. The *Piecewise constant hazards* model is treated in the functions pch (individual data) and tpch (tabular data).

The *lognormal* and *loglogistic* distributions are also included as possible choices and allow for hazard functions that are first increasing to a maximum and then decreasing, while the other distributions all have monotone hazard functions. However, since these families are not closed under proportional hazards without artificially adding a third, "proportionality", parameter, they are not discussed here (regard these possibilities as experimental). It is better to combine the lognormal and the loglogistic distributions with the accelerated failure time modeling, where they naturally fit in.

See Figure 8.1 for Weibull and Gompertz hazard functions with selected parameter values.

We note in passing that the fourth case, the Gompertz model with negative rate parameter, does not represent a true survival distribution, because the hazard function decreases too fast: There will be a positive probability of eternal life.

Experience shows that the Gompertz distribution fits adult mortality very well, in the ages 30 to 85, say. The modeling of mortality from birth to early adulthood, on the other hand, is demanding since the typical hazard function for all these ages is U-shaped with high infant mortality and relatively low child mortality. Since both the Weibull and the Gompertz distributions have a monotone hazard function, neither is suitable to fit the mortality of the *full*

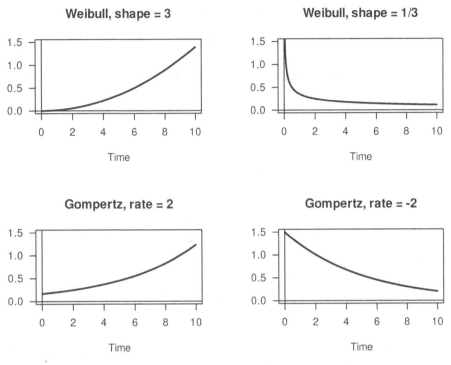

**FIGURE 8.1** Selected hazard functions.

human life span. However, both distributions are suitable for fitting shorter pieces of the life span, and for longer spans there are two possibilities, a nonparametric model (Cox regression) and a piecewise constant hazard model, where the former can be seen as a limiting case of the latter. More about that later.

### 8.1.1   The Weibull model

The family of *Weibull* distributions may be defined by the family of hazard functions

$$h(t; p, \lambda) = \frac{p}{\lambda}\left(\frac{t}{\lambda}\right)^{p-1}, \quad t, p, \lambda > 0. \tag{8.1}$$

If we start with

$$h_1(t) = t^{p-1}, \quad p \geq 0, \, t > 0$$

for a *fixed* $p$, and generate a proportional hazards family from there,

$$h_c(t) = ch_1(t), \quad c, t > 0,$$

we get

$$h_c(t) = ct^{p-1} = \frac{p}{\lambda}\left(\frac{t}{\lambda}\right)^{p-1} \tag{8.2}$$

by setting

$$c = \frac{p}{\lambda^p},$$

which shows that for each fixed $p > 0$, a proportional hazards family is generated by varying $\lambda$ in equation (8.2). On the other hand, if we pick two members from the family equation (8.2) with *different* values of $p$, they would not be proportional.

To summarize, the *Weibull* family of distributions is not *one* family of proportional hazards distributions, but a *collection* of families of proportional hazards. The collection is indexed by $p > 0$. It is true, though, that all the families are closed under the Weibull distribution.

The proportional hazards regression model with a Weibull baseline distribution is obtained by multiplying (8.1) by $\exp(\beta x)$:

$$h(t; x, \lambda, p, \beta) = \frac{p}{\lambda}\left(\frac{t}{\lambda}\right)^{p-1} e^{\beta x}, \quad t > 0. \tag{8.3}$$

The function **phreg** in package **eha** fits models like equation (8.3) by default.

### 8.1.2 The Gompertz distribution

The Gompertz families of distributions are defined in essentially two ways in the **R** package eha: The *rate* and the *canonical* representations. The reason for this duality is that the families need to be differently represented depending on whether proportional hazards or accelerated failure time models are under consideration.

In the *proportional hazards* case, the *rate* formulation is used, and it is characterized by an exponentially increasing hazard function with fixed rate r:

$$h(t; p, r) = pe^{rt}, \quad p, t > 0; -\infty < r < \infty. \tag{8.4}$$

As noted earlier, when $r < 0$, the hazard function $h$ is decreasing "too fast" to define a proper survival function, and $r = 0$ gives the *exponential distribution* as a special case. And for each fixed $r$, the family of distributions indexed by $p > 0$ constitutes a proportional hazards family of distributions, and the corresponding regression model is written as

$$h(t; x, p, r, \beta) = pe^{rt}e^{\beta x}, \quad t > 0. \tag{8.5}$$

### 8.1.3 Application

The data set **oldmort** in the **R** package **eha** contains life histories of people followed from their 60th birthday to their 100th, or until death, born between June 28, 1765 and December 31, 1820 in Skellefteå. The data set is described in detail in Chapter 1. The variable enter is age at start of the given interval, and exit contains the age at the end of the interval. We need to calculate *follow-up time* since age 60, so a new data frame, olm, is created as a copy of oldmort, and then 60 is subtracted from enter and exit. See the result in Table 8.1, where the most relevant variables for our purpose are shown.

The variable names are more or less self explanatory, *enter* and *exit* are time in years since the sixtieth birthday, *age* is age at start

**TABLE 8.1** The data set 'olm', first rows.

| birthdate | sex | region | enter | exit | event | age |
|---|---|---|---|---|---|---|
| 1765-06-28 | female | rural | 34.510 | 35.813 | TRUE | 94.510 |
| 1765-09-25 | female | industry | 34.266 | 35.756 | TRUE | 94.266 |
| 1768-11-27 | female | rural | 31.093 | 31.947 | TRUE | 91.093 |
| 1770-12-28 | female | industry | 29.009 | 29.593 | TRUE | 89.009 |
| 1770-01-01 | female | rural | 29.998 | 30.211 | TRUE | 89.998 |

of follow-up (the original `enter` variable). The variable *event* is an indicator of death at the duration given by `exit`.

To have something to compare to, a Cox regression is performed first, see Table 8.2. Two covariates are included in the model, `sex` and `region`. Both are *categorical*, `region` with three categories, *town* (reference), *industry*, and *rural*, and `sex` with *male* as reference category.

**TABLE 8.2** Old age mortality, Cox proportional hazards model.

| Covariate | | Mean | Coef | Rel.Risk | S.E. | L-R p |
|---|---|---|---|---|---|---|
| sex | | | | | | 0.000 |
| | *male* | 0.406 | 0 | 1 | (reference) | |
| | *female* | 0.594 | −0.187 | 0.830 | 0.046 | |
| region | | | | | | 0.001 |
| | *town* | 0.111 | 0 | 1 | (reference) | |
| | *industry* | 0.326 | 0.212 | 1.236 | 0.085 | |
| | *rural* | 0.563 | 0.052 | 1.053 | 0.083 | |
| Events | | 1971 | TTR | 37824 | | |
| Max. Log Likelihood | | −13563 | | | | |

Then a Weibull model is fitted, see Table 8.3.

```
fit <- phreg(Surv(enter - 60, exit - 60, event) ~ sex + region,
        dist = "weibull", data = oldmort)
```

A closer look at the estimates of regression coefficients shows that

**TABLE 8.3** Old age mortality, Weibull proportional hazards model.

| Covariate | | Mean | Coef | Rel.Risk | S.E. | L-R p |
|---|---|---|---|---|---|---|
| sex | | | | | | 0.001 |
| | *male* | 0.406 | 0 | 1 | (reference) | |
| | *female* | 0.594 | −0.159 | 0.853 | 0.046 | |
| region | | | | | | 0.000 |
| | *town* | 0.111 | 0 | 1 | (reference) | |
| | *industry* | 0.326 | 0.249 | 1.283 | 0.085 | |
| | *rural* | 0.563 | 0.052 | 1.054 | 0.083 | |
| Events | | 1971 | TTR | 37824 | | |
| Max. Log Likelihood | | −7421 | | | | |

**TABLE 8.4** Coefficients with Cox and Weibull regressions, data oldmort.

| | Cox | Weibull |
|---|---|---|
| sexfemale | −0.1867 | −0.1587 |
| regionindustry | 0.2120 | 0.2491 |
| regionrural | 0.0516 | 0.0525 |

they are not very close, in Table 8.4 they are put side by side for easier comparison.

Let us compare the estimated cumulative baseline hazard functions, see Figure 8.2.

This is not a good fit, it seems as if the Weibull hazard cannot grow fast enough. A better approach is to fit a *Gompertz* distribution, and check parameter and baseline hazards estimates, see Figure 8.3 and Table 8.6.

The Gompertz model fits the baseline hazard very well up until duration 30 (age 90), but after that the exponential growth slows down. The early growth rate is 9.5 per cent per year.

The result of fitting the Gompertz model is shown in Table 8.5.

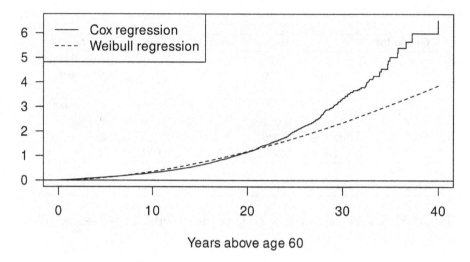

**FIGURE 8.2** Baseline cumulative hazards for Cox and Weibull regressions.

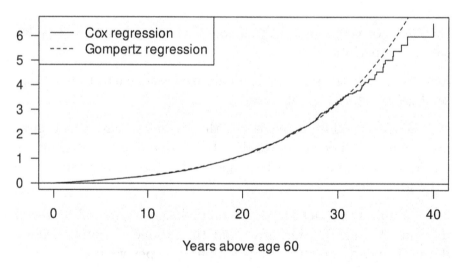

**FIGURE 8.3** Baseline cumulative hazards for Cox and Gompertz regressions.

**TABLE 8.5** Old age mortality, Gompertz PH model.

| Covariate | | Mean | Coef | Rel.Risk | S.E. | L-R p |
|---|---|---|---|---|---|---|
| sex | | | | | | 0.000 |
| | *male* | 0.406 | 0 | 1 | (reference) | |
| | *female* | 0.594 | −0.190 | 0.827 | 0.046 | |
| region | | | | | | 0.002 |
| | *town* | 0.111 | 0 | 1 | (reference) | |
| | *industry* | 0.326 | 0.207 | 1.230 | 0.085 | |
| | *rural* | 0.563 | 0.047 | 1.048 | 0.083 | |
| Events | | 1971 | TTR | 37824 | | |
| Max. Log Likelihood | | −7281 | | | | |

**TABLE 8.6** Coefficients with Gompertz, Cox and Weibull regressions, time scale duration since age 60.

| | Gompertz | Cox | Weibull |
|---|---|---|---|
| sexfemale | −0.190 | −0.187 | −0.159 |
| regionindustry | 0.207 | 0.212 | 0.249 |
| regionrural | 0.047 | 0.052 | 0.052 |

The Gompertz and Cox models are very close, both regarding regression parameter estimates and baseline hazard functions.

### 8.1.4 The parametric model with left truncation

The data set oldmort contains left-truncated life histories as a consequence of using *age* as time scale. In the presentation above we chose to change the time scale so that the origin was age 60 (sharp). This is of no importance when fitting the semi-parametric Cox regression, an additive change of time scale will only shift the estimated cumulative hazards along the x-axis.

But for parametric models it matters, and we illustrate it by repeating the previous Weibull, Gompertz, and Cox regression analyses with the original time scale. The consequence is that focus is shifted from a conditional (on survival until 60) analysis to an

**TABLE 8.7** Coefficients with Gompertz, Cox, and Weibull regressions, time scale age.

|                | Gompertz | Cox | Weibull |
|----------------|----------|-----|---------|
| sexfemale      | −0.190   | −0.187 | −0.187 |
| regionindustry | 0.207    | 0.212  | 0.209  |
| regionrural    | 0.047    | 0.052  | 0.045  |

unconditional, where the baseline hazard and regression coefficients are estimated for the *full life span* (0–100 years of age).

Compare these fitted coefficients with the earlier from Table 8.6. For the Gompertz and Cox regression models, coefficients are identical, while they differ for the Weibull distribution.

Let us look at the baseline cumulative hazards, see Figure 8.4, where the graph is cut at ages 60 and 80 for clarity. Notice the value at Time = 60: The parametric models are "extrapolating" back to time at birth, and so the estimates do not represent the cumulative hazards of the conditional distribution, given survival to age 60. This has an impact on the estimates of the regression coefficients in the case of the Weibull distribution, because the *conditional* is *not* Weibull even though the unconditional is. This phenomenon does *not* apply to the Gompertz distribution, for which the conditional distribution is again Gompertz with the same rate, but with a different level.

From Figure 8.4 it appears as if we can get the *conditional* cumulative hazards simply by subtracting the value at age 60, $H(60)$, from the whole curves, and that is in fact correct. In the Gompertz case, it would simply recover Figure 8.3, but the Weibull case is different: Starting with Figure 8.2 and adding the adjusted Weibull curve from Figure 8.4 we get Figure 8.5.

Obviously, the conditional Weibull distribution fits data much better than the unconditional one. The comparison with the Gompertz distribution is shown in Figure 8.6.

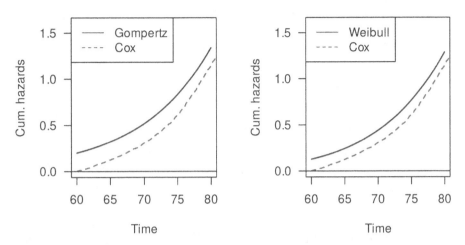

**FIGURE 8.4** Gompertz vs Cox and Weibull vs Cox estimated cumulative hazards functions.

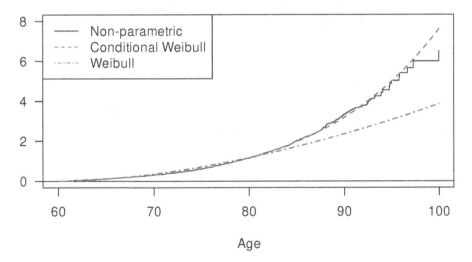

**FIGURE 8.5** Comparison of the conditional and unconditional Weibull models.

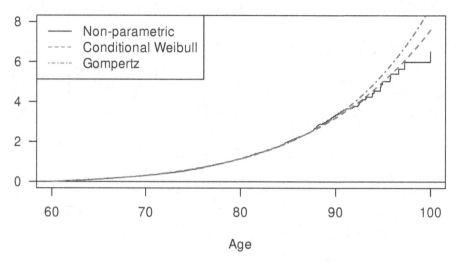

**FIGURE 8.6** Comparison of the Gompertz and conditional Weibull models.

It seems as if the conditional Weibull model fits data as good as or even better than the Gompertz model. The latter grows too fast in the very high ages, and this is an observation found in many studies lately (Rootzén and Zholud, 2017; Lenart and Vaupel, 2017; Barbi et al., 2018; Broström, 2019).

### 8.1.5   The piecewise constant proportional hazards model

The *piecewise constant* hazard (pch) model can always be used in the fitting process with good results. It is however an uncertainty moment in the process: How should time be cut into pieces, and how many pieces should there be? Two possible strategies, (i) choose equally-spaced cut points, and (ii) relatively more cut points where there are many deaths, that is, where the hazard function is expected to be steep.

The oldmort data set spans a time interval of length 40 years, and we know that mortality on the age interval 60–100 is increasing almost exponentially, suggesting more cut points in the high ages. Against that is the fact that in very high ages, say 90–100, not many observations are still around, most of them have already died.

We may start with eight intervals of equal length, 60–65, 65–70, ...,
95–100, and fit a pch model with the aid of the function `pchreg` in
`eha`. The result is shown in Table 8.8.

**TABLE 8.8** Old age mortality, pch proportional hazards model.

| Covariate | | Mean | Coef | Rel.Risk | S.E. | L-R p |
|---|---|---|---|---|---|---|
| sex | | | | | | 0.000 |
| | *male* | 0.406 | 0 | 1 | (reference) | |
| | *female* | 0.594 | −0.182 | 0.834 | 0.046 | |
| region | | | | | | 0.001 |
| | *town* | 0.111 | 0 | 1 | (reference) | |
| | *industry* | 0.326 | 0.223 | 1.250 | 0.085 | |
| | *rural* | 0.563 | 0.059 | 1.060 | 0.083 | |
| Events | | 1971 | TTR | 37824 | | |
| Max. Log Likelihood | | −7295 | | | | |

Then the baseline cumulative hazards are compared to the one
from the Cox regression fit in Figure 8.7.

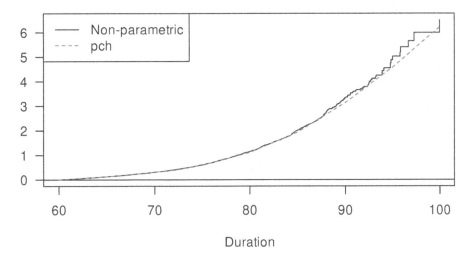

**FIGURE 8.7** Piecewise constant cumulative hazards, old age
mortality.

As expected, a good fit. It is also obvious that the estimated regression parameters do not vary much between the studied models.

**TABLE 8.9** Table version of the data frame 'oldmort'.

| sex    | region   | age   | event | exposure |
|--------|----------|-------|-------|----------|
| male   | town     | 60—65 | 29    | 623.060  |
| female | town     | 60—65 | 23    | 1230.859 |
| male   | industry | 60—65 | 71    | 1946.578 |
| female | industry | 60—65 | 67    | 2481.924 |
| male   | rural    | 60—65 | 83    | 3456.088 |

The piecewise constant model works well with this data set, but its real strength is its flexibility and speed with huge data sets. Its full potential is maximized by initially tabulate the data by using the eha function toTpch. An illustration with the oldmort data set.

```
olmtab <- toTpch(Surv(enter, exit, event) ~ sex + region,
                cuts = c(seq(60, 85, by = 5), 100), data = oldmort)
```

The resulting table (first five rows) is shown in Table 8.11.

The original data set has 6495 rows (observations), while the created table has only 36 rows. The latter is analyzed via the function tpchreg, see Table 8.10, which is identical to Table 8.8.

```
fit.tpch <- tpchreg(oe(event, exposure) ~ sex + region,
                    data = olmtab, time = age)
```

### 8.1.6   Testing the proportional hazards assumption

The pch model is well suited for a formal test of the proportional hazards model, but some trickery is needed with the eha package. It is best shown by example, and we continue by utilizing the newly created data table olmtab.

**TABLE 8.10** Proportional hazards with table version of 'oldmort'.

| Covariate | | Mean | Coef | Rel.Risk | S.E. | L-R p |
|---|---|---|---|---|---|---|
| sex | | | | | | 0.000 |
| | *male* | 0.406 | 0 | 1 | (reference) | |
| | *female* | 0.594 | −0.181 | 0.834 | 0.046 | |
| region | | | | | | 0.001 |
| | *town* | 0.111 | 0 | 1 | (reference) | |
| | *industry* | 0.326 | 0.224 | 1.251 | 0.085 | |
| | *rural* | 0.563 | 0.060 | 1.062 | 0.083 | |
| Events | | 1971 | TTR | 37824 | | |
| Max. Log Likelihood | | −7296 | | | | |
| Restricted mean survival: | | 15.8 | in (60, 100] | | | |

By omitting the `time` argument in the call to `tpchreg`, an exponential (constant hazards) model is fitted, and the variable `age` is free to be included as a covariate.

```
fit.tpch <- tpchreg(oe(event, exposure) ~ age * (sex + region),
                data = olmtab)
(dr <- drop1(fit.tpch, test = "Chisq"))
```

```
Single term deletions

Model:
oe(event, exposure) ~ age * (sex + region)
           Df   AIC     LRT Pr(>Chi)
<none>          14614
age:sex     5 14610   7.009  0.21997
age:region 10 14614  20.108  0.02825
```

The age interaction with `sex` is very non-significant, while the effect of region on mortality seems to vary significantly with age. It can be illustrated by performing a stratified (by region) analysis with the `time = age` variable included in the usual way, then plotting the hazard functions for the three strata, see Figure 8.8.

```
fit.str <- tpchreg(oe(event, exposure) ~ sex + strata(region),
                   time = age, data = olmtab)
```

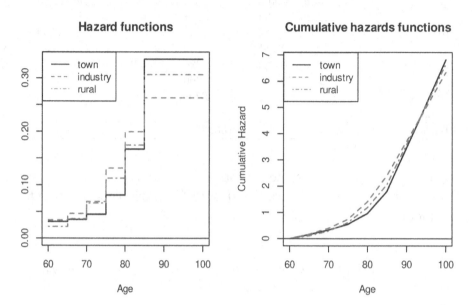

**FIGURE 8.8** Hazard functions for three regions, old age mortality, Skellefteå 1860–1880.

The deviating region is town. It is also the smallest region, with no registered deaths above age 90 with only 4.7 person years. We also note that mortality in ages 85–90 is highest in the town, while in the younger ages the town region has the lowest mortality. The zero mortality in ages above age 90 is simply an artifact depending on very few observed person years.

Let us perform the same exercise with a larger data set, the Swedish population 1968–2019. (See also Example 3.1 in Chapter 3.) We check the hypothesis of proportional hazards between women and men, the question is: Is the female advantage of the same relative size in all ages?

```
sp <- swepop
sp$deaths <- swedeaths$deaths
```

Then

```
fit.swr <- tpchreg(oe(deaths, pop) ~ strata(sex) + I(year - 2000),
                last = 101,
                time = age, data = sp)
rr.sex <- exp(tpchreg(oe(deaths, pop) ~ sex + I(year - 2000),
                last = 101,
                time = age, data = sp)$coefficients[1])
cumhaz <- hazards(fit.swr, cum = TRUE) # Cumulative hazards
haz <- hazards(fit.swr, cum = FALSE) # NOT Cumulative hazards
```

```
op <- par(mfrow = c(1, 2))
plot(haz$x, haz$y[2, ] / haz$y[1, ], type = "l", ylim = c(1, 3),
     xlab = "Age", ylab = "Hazard Ratio")
abline(h = 1)
abline(h = rr.sex, lty = 2)
text(5, 1.65, "PH")
plot(cumhaz$x, cumhaz$y[2, ] / cumhaz$y[1, ], type = "l",
     ylim = c(1, 3),
     xlab = "Age", ylab = "Cumulative Hazard Ratio")
abline(h = 1)
abline(h = rr.sex, lty = 2)
text(5, 1.65, "PH")
```

```
par(op)
```

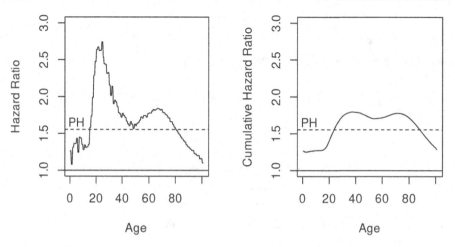

**FIGURE 8.9** Mortality ratio for men vs. women, Sweden 1968–2019.

We note two things: (i) The variation around the proportional hazards estimate (PH) is huge and (ii) the smoothing effect of accumulation is large, which we should keep in mind as a warning when trying to judge proportionality in graphs of cumulative hazards functions.

### 8.1.7   Choosing the best parametric proportional hazards model

For modeling survival data with parametric proportional hazards models, the distributions of the function `phreg` in the package **eha** are available. How to select a suitable parametric model is shown by a couple of examples using the now familiar data set `oldmort`.

Remember that by design, individuals are followed from the day they are aged 60. In order to calculate the follow-up times, we usually subtract 60 from the two columns enter and exit. Otherwise, when specifying a parametric survivor distribution, it would in fact correspond to a left-truncated (at 60) distribution. However, for a Cox regression, this makes no difference.

**TABLE 8.11** Comparison of conditional and unconditional maximum likelihood values.

| Weibull | Gompertz | EV | PCH |
|---|---|---|---|
| 14572.91 | 14571.74 | 14586.58 | 14609.55 |
| 14852.22 | 14571.74 | 14586.58 | 14609.55 |

```
om <- oldmort
fm <- as.formula("Surv(enter, exit, event) ~ sex + region")
fm0 <- as.formula("Surv(enter - 60, exit - 60, event) ~ sex + region")
fit.w <- phreg(fm, data = oldmort)
o.w <- extractAIC(fit.w)[2]
fit.w0 <- phreg(fm0, data = oldmort)
o.w0 <- extractAIC(fit.w0)[2]
```

Here we applied a *Weibull* baseline distribution (the *default* distribution in phreg; by specifying nothing, the Weibull is chosen). Note also the way we can store a formula for future use: This is particularly useful when the same model will be fitted several times while changing some attribute, like baseline distribution. Now let us repeat this with the pch, gompertz and ev distributions in the phreg package, for both the unconditional and conditional approaches.

Then we compare the AICs and choose the distribution with the smallest value.

First we note that only for the Weibull distribution it makes a difference which time scale (age or duration) is used. The Gompertz distribution gives the best fit, but the *conditional* Weibull (age as time scale) is very close. We also saw this graphically earlier (Figure 8.6).

The reason that the pch model fares so badly is that it is punished for the large number of parameters (eight) it uses to estimate the baseline hazard.

## 8.2   Accelerated Failure Time Models

The accelerated failure time (AFT) model is best described through
relations between survivor functions. For instance, comparing two
groups:

- **Group 0:** $P(T \geq t) = S_0(t)$ (control group)
- **Group 1:** $P(T \geq t) = S_0(\phi t)$ (treatment group)

The model says that treatment *accelerates* failure time by the
factor $\phi$. If $\phi < 1$, treatment is good (prolongs life), otherwise bad.
Another interpretation is that the *median* life length is *multiplied*
by $1/\phi$ by treatment.

In Figure 8.10 the difference between the accelerated failure time
and the proportional hazards models concerning the hazard functions is illustrated.

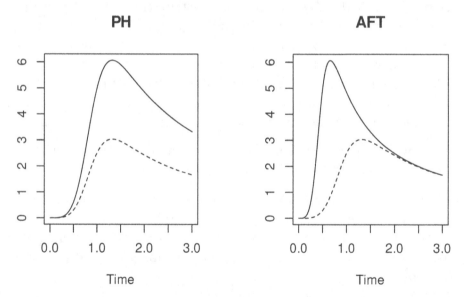

**FIGURE 8.10** Proportional hazards (left) and accelerated failure
time model (right). The baseline distribution is Loglogistic with
shape 5 (dashed).

The AFT hazard is not only multiplied by 2, it is also shifted to the left; the process time is accelerated. Note how the hazards in the AFT case converges as time increases. This is usually a sign of the suitability of an AFT model.

### 8.2.1 The AFT regression model

If $T$ has survivor function $S(t)$ and $T_c = T/c$, then $T_c$ has survivor function $S(ct)$. Then, if $Y = \log(T)$ and $Y_c = \log(T_c)$, the following relation holds:

$$Y_c = Y - log(c).$$

With $Y = \epsilon$, $Y_c = Y$, and $\log(c) = -\beta\mathbf{x}$ this can be written in familiar form:

$$Y = \beta\mathbf{x} + \epsilon,$$

i.e., an ordinary linear regression model for the log survival times. In the absence of right censoring and left truncation, this model can be estimated by least squares. However, the presence of these forms of incomplete data makes it necessary to rely on maximum likelihood methods. In **R**, the functions `aftreg` in the package **eha** and the function `survreg` in the package **survival** that perform the task of fitting AFT models. The package **flexsurv** (Jackson, 2016) has some useful functionality in this area.

Besides differing parameterizations, the main difference between `aftreg` and `survreg` is that the latter does not allow for left truncated data. One reason for this is that left truncation is a much harder problem to deal with in AFT models than in proportional hazards models. The reason is that, with a time varying covariate $z(t), t \geq 0$, the AFT model is

$$S(t; z) = S_0\left(\int_0^t \exp(\beta z(s))ds\right),$$

and it is required that $z(s)$ is known for all $s, 0 \leq s \leq t$. With a left truncated observation at $t_0$, say, $z(s)$ is unknown for $0 \leq s < t_0$. In eha, this is solved by *assuming* that $z(s) = z(t_0), 0 \leq s < t_0$.

A detailed description of the implementation of the AFT models in eha is found in Appendix B.

### 8.2.2   AFT modeling in R

We repeat the examples from the proportional hazards section, but with AFT models instead.

**Example 8.1** (Old age mortality)

For a description of this data set, see above. Here we fit an AFT model with the *Weibull* distribution. This should be compared to the proportional hazards model with the Weibull distribution, see Table 8.3.

```
source("R/fit.out.R")
fm <- as.formula("Surv(enter - 60, exit - 60, event) ~ sex + region")
fit.w1 <- aftreg(fm, id = id, data = oldmort)
```

Note carefully the inclusion of the argument id, it is necessary when some individuals are represented by more than one record in the data, the results are shown in Table 8.12.

Note that the "Max. log. likelihood", $-7415$, is not exactly the same, and the reason is the presence of time-varying covariate region. With no time-varying covariates, the AFT and the PH models are equivalent. $\square$

**Example 8.2** (Length of birth intervals.)

The data set **fert** in the **R** package **eha** contains birth intervals for married women in 19th century Skellefteå. It is described in detail in Chapter 1. Here only the intervals starting with the first

**TABLE 8.12** Old age mortality in 19th century Skellefteå, AFT model.

| Covariate | | Mean | Coef | Acc'd time | S.E. | L-R $p$ |
|---|---|---|---|---|---|---|
| sex | | | | | | 0.001 |
| | *male* | 0.406 | 0 | 1 | (reference) | |
| | *female* | 0.594 | −0.092 | 0.912 | 0.027 | |
| region | | | | | | 0.000 |
| | *town* | 0.111 | 0 | 1 | (reference) | |
| | *industry* | 0.326 | 0.151 | 1.163 | 0.050 | |
| | *·rural* | 0.563 | 0.032 | 1.032 | 0.048 | |
| Baseline parameters | | | | | | |
| log(scale) | | | 2.903 | 18.227 | 0.048 | 0.000 |
| log(shape) | | | 0.545 | 1.724 | 0.018 | 0.000 |
| Baseline expected life: | | | | | | |
| Events | | 1971 | TTR | 37824 | | |
| Max. Log Likelihood | | −7415 | $p$-value | 0 | | |

**TABLE 8.13** The fertility data set, first few rows.

| id | parity | age | year | next.ivl | event | prev.ivl | ses | parish |
|---|---|---|---|---|---|---|---|---|
| 1 | 1 | 25 | 1826 | 8.000 | 0 | 0.411 | farmer | SKL |
| 2 | 1 | 19 | 1821 | 1.837 | 1 | 0.304 | unknown | SKL |
| 3 | 1 | 24 | 1827 | 2.051 | 1 | 0.772 | farmer | SKL |
| 4 | 1 | 35 | 1838 | 1.782 | 0 | 6.787 | unknown | SKL |
| 5 | 1 | 28 | 1832 | 1.629 | 1 | 3.031 | farmer | SKL |
| 6 | 1 | 25 | 1829 | 1.730 | 1 | 0.819 | lower | SKL |

birth for each woman are considered. First the data are extracted and examined, see Table 8.13.

Some women never got a second child, for instance the first woman (**id = 1**) above.

It turns out that, as is reasonably expected, that the hazard function for time to the second birth is first increasing to a maximum, then decreasing. Indeed, we can make a crude check of that by estimating the the hazard function without covariates with the piecewise constant hazard model, which in principle is a non-parametric approach if the time span is cut in small enough pieces. So we cut the first eight years in half-year-long pieces, see Figure 8.11.

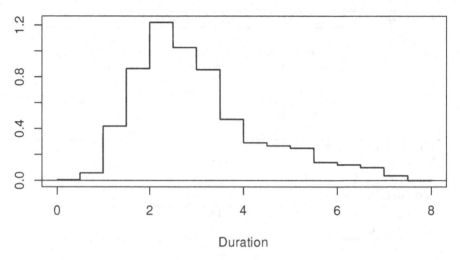

Duration

**FIGURE 8.11** Estimated hazard function for waiting time between first and second birth.

There are two possible candidates for the baseline distribution that has the right shape of the hazard function: The *Lognormal* and the *Loglogistic* distributions. □

### 8.2.3   The Lognormal model

The family of *lognormal* distributions is characterized by the fact that taking the natural logarithm of a random variable from the family gives a random variable from the family of *Normal* distributions. Both the hazard and the survivor functions lack closed forms.

The data are already extracted and examined in the previous example. Now, the accelerated failure time (AFT) *lognormal* regression model is examined.

```
fit.lognorm <- aftreg(Surv(next.ivl, event) ~ age + I(year - 1860) + ses
                data = f12, dist = "lognormal")
```

Note that the argument id is *not* used here, that is because there are no time varying covariates for next.ivl. The id variable in the data set refers to *mother's id* and not to specific birth intervals. The data set f12 is a subset of fert, it only includes intervals starting with the first birth.

**TABLE 8.14** Length of birth intervals, Lognormal AFT model.

| Covariate | | Mean | Coef | Acc'd time | S.E. | L-R $p$ |
|---|---|---|---|---|---|---|
| age | | 27.247 | −0.018 | 0.982 | 0.003 | 0.000 |
| I(year − 1860) | | −1.041 | 0.000 | 1.000 | 0.001 | 0.996 |
| ses | | | | | | 0.122 |
| | *farmer* | 0.458 | 0 | 1 | (reference) | |
| | *unknown* | 0.178 | −0.063 | 0.939 | 0.034 | |
| | *upper* | 0.022 | 0.038 | 1.038 | 0.082 | |
| | *lower* | 0.341 | −0.049 | 0.953 | 0.027 | |
| Baseline parameters | | | | | | |
| log(scale) | | | 0.237 | 1.268 | 0.076 | 0.002 |
| log(shape) | | | 0.700 | 2.013 | 0.018 | 0.000 |
| Baseline expected life: | | | | | | |
| Events | | 1646 | TTR | 4156 | | |
| Max. Log Likelihood | | −2426 | $p$-value | 0 | | |

The interpretation of the regression coefficients is different from the PH case: Exponentiated (*lifeAcc*) they measure how much time to event is *accelerated*. For instance, if mother's age is increased by one year, the waiting time to the next birth is accelerated by the factor 0.9822, that is, slowed down slightly.

It may be more natural to use *expected life* as a comparison rather than *accelerated time*, even if one is the reverse of the other. In aftreg, it is possible to choose param = lifeExp for that puropse.

```
fit.lognorm2 <- aftreg(Surv(next.ivl, event) ~ age +
                    I(year - 1860) + ses,
                data = f12, dist = "lognormal",
                param = "lifeExp")
```

**TABLE 8.15** Length of birth intervals, Lognormal AFT model, reverse parameterization.

| Covariate | | Mean | Coef | Ext'd life | S.E. | L-R *p* |
|---|---|---|---|---|---|---|
| age | | 27.247 | 0.018 | 1.018 | 0.003 | 0.000 |
| I(year - 1860) | | -1.041 | -0.000 | 1.000 | 0.001 | 0.996 |
| ses | | | | | | 0.122 |
| | *farmer* | 0.458 | 0 | 1 | (reference) | |
| | *unknown* | 0.178 | 0.063 | 1.065 | 0.034 | |
| | *upper* | 0.022 | −0.038 | 0.963 | 0.082 | |
| | *lower* | 0.341 | 0.049 | 1.050 | 0.027 | |
| Baseline parameters | | | | | | |
| log(scale) | | | 0.237 | 1.268 | 0.076 | 0.002 |
| log(shape) | | | 0.700 | 2.013 | 0.018 | 0.000 |
| Baseline expected life: | | | | | | |
| Events | | 1646 | TTR | 4156 | | |
| Max. Log Likelihood | | −2426 | *p*-value | 0 | | |

As can be seen by comparing Tables 8.14 and, 8.15 the only difference is the signs of the estimated coefficients. And therefore, "lifeAcc = 1 / lifeExp". The simple interpretation of the "lifeExp" parameter is that it multiplies the expected value for the corresponding category. In Figure 8.12 the estimated baseline hazard function is shown.

It was created using

```
plot(fit.lognorm, fn = "haz", main = "",
    xlab = "Years", ylab = "Hazards")
```

□

### 8.2.4   The Loglogistic model

The family of *loglogistic* distributions is characterized by the fact that taking the natural logarithm of a random variable from the

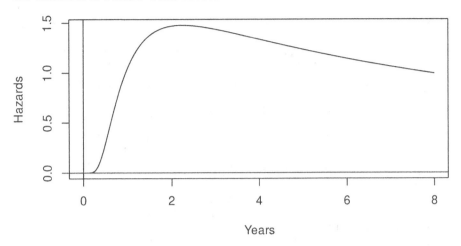

**FIGURE 8.12** Estimated lognormal baseline hazard function for length of birth intervals.

family gives a random variable from the family of *Logistic* distributions. Contrary to the Lognormal case, both the hazard and the survivor functions have closed forms.

The data are already extracted and examined in the previous example. Now, the accelerated failure time (AFT) *loglogistic* regression model is examined.

```
fit.loglogist <- aftreg(Surv(next.ivl, event) ~ age +
                  I(year - 1860) + ses,
              data = f12, dist = "loglogistic")
```

Which is the better fit, the lognormal or the loglogistic? Let us compare the maximized log likelihoods: For the lognormal it is –2426.23, and for the loglogistic it is –2292.01, so it seems as if the *loglogistic* fit is clearly better. Note, though, that this is *not* a formal statistical test, it is a comparison by the *Akaike* criterion (Akaike, 1974). This conclusion is also partly supported by a comparison of the two estimated baseline hazard functions (Figures 8.13 and 8.12) with the estimated crude hazard function in Figure 8.11.

**TABLE 8.16** Length of birth intervals, Loglogistic AFT model.

| Covariate | | Mean | Coef | Acc'd time | S.E. | L-R *p* |
|---|---|---|---|---|---|---|
| age | | 27.247 | −0.014 | 0.986 | 0.002 | 0.000 |
| I(year − 1860) | | −1.041 | 0.000 | 1.000 | 0.001 | 0.819 |
| ses | | | | | | 0.206 |
| | *farmer* | 0.458 | 0 | 1 | (reference) | |
| | *unknown* | 0.178 | −0.029 | 0.971 | 0.029 | |
| | *upper* | 0.022 | 0.098 | 1.103 | 0.075 | |
| | *lower* | 0.341 | −0.033 | 0.968 | 0.024 | |
| Baseline parameters | | | | | | |
| log(scale) | | | 0.327 | 1.386 | 0.068 | 0.000 |
| log(shape) | | | 1.366 | 3.918 | 0.021 | 0.000 |
| Baseline expected life: | | | | | | |
| Events | | 1646 | TTR | 4156 | | |
| Max. Log Likelihood | | −2292 | *p*-value | 0 | | |

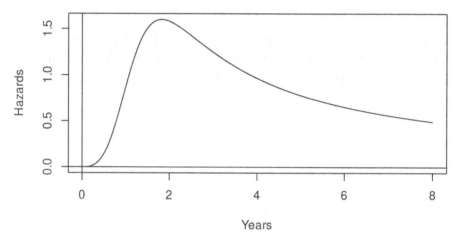

**FIGURE 8.13** Estimated loglogistic baseline hazard function for length of birth intervals.

### 8.2.5   The Gompertz model

The *Gompertz* distribution is special in that it can be fit into both the AFT and the PH framework. Of course, as we have seen, this also holds for the Weibull distribution in a trivial way, the AFT and the PH models are the same, but for the Gompertz distribution, the AFT and PH approaches yield different models.

For the AFT framework to be in place in the Gompertz case, it needs to be formulated with a rather unfamiliar parameterization, which is called *the canonical parameterization* in the package eha. It works as follows. The standard definition of the Gompertz hazard function is

$$h_r(t; (\alpha, \beta)) = \alpha \exp(\beta t), \quad t > 0; \ \alpha > 0, -\infty < \sigma < \infty.$$

and it is called the *rate* parameterization in eha. As noted earlier, in order for $h_r$ to determine a proper survival distribution, it must be required that $\beta \geq 0$. It was also noted that when $\beta = 0$, the resulting distribution is *exponential*.

The *canonical* definition of the Gompertz hazard function is given by

$$h_c(t; (\tau, \sigma)) = \frac{\tau}{\sigma} \exp\left(\frac{t}{\sigma}\right), \quad t > 0; \ \tau, \sigma > 0.$$

The transition from $h_r$ to $h_c$ is given by $\sigma = 1/\beta$, $\tau = \alpha/\beta$, and note that this implies that the rate in the canonical form must be strictly positive. Furthermore, the exponential special case now appears in the limit as $\sigma \to \infty$. In practice this means that when the baseline hazard is only weakly increasing, $\sigma$ is very large and numerical problems in the estimation process is likely to occur.

The conclusion of all this is that the AFT Gompertz model is suitable in situations where the intensity of an event is clearly increasing with time. A good example is adult mortality.

We repeat the PH analysis in Table 8.5, but with the AFT model, see Table 8.17.

The expected remaining life at 60 for a man living in the town is *16 years*, and for a woman living in the town the expected remaining life is $1.067 \times 16 = 17$ years.

**TABLE 8.17** Old age mortality (60–100), Gompertz AFT model.

| Covariate | | Mean | Coef | Ext'd life | S.E. | L-R $p$ |
|---|---|---|---|---|---|---|
| sex | | | | | | 0.001 |
| | male | 0.406 | 0 | 1 | (reference) | |
| | female | 0.594 | 0.065 | 1.067 | 0.019 | |
| region | | | | | | 0.007 |
| | town | 0.111 | 0 | 1 | (reference) | |
| | industry | 0.326 | -0.096 | 0.908 | 0.039 | |
| | rural | 0.563 | -0.046 | 0.955 | 0.039 | |
| Baseline parameters | | | | | | |
| log(scale) | | | 2.380 | 10.802 | 0.050 | 0.000 |
| log(shape) | | | -1.604 | 0.201 | 0.072 | 0.000 |
| Baseline expected life: | | 16 | | | | |
| Events | | 1971 | TTR | 37824 | | |
| Max. Log Likelihood | | -7288 | $p$-value | 6.9e-05 | | |

## 8.3   Proportional Hazards or AFT Model?

The problem of choosing between a proportional hazards and an accelerated failure time model (everything else equal) can be solved by comparing the AIC (de Leeuw, 1992; Akaike, 1974) of the models. Since the numbers of parameters are equal in the two cases, this amounts to comparing the maximized likelihoods. For instance, in the case with *old age mortality*:

Let us see what happens with the *Gompertz* AFT model: Exactly the same procedure as with the *Weibull* distribution, but we have to specify the Gompertz distribution in the call (remember, the *Weibull* distribution is the default choice, both for phreg and aftreg).

Comparing the corresponding result for the proportional hazards and the AFT models with the Gompertz distribution, we find that the maximized log likelihood in the former case is –7280.9, compared to –7288.3 for the latter. This indicates that the proportional hazards model fit is better. Note however that we cannot formally test the proportional hazards hypothesis; the two models are not nested.

---

## 8.4 Discrete Time Models

There are two ways of looking at discrete duration data; either time is truly discrete, for instance the number of trials until an event occurs, or an approximation due to rounding of continuous time data. In a sense all data are discrete, because it is impossible to measure anything on a continuous scale with infinite precision, but from a practical point of view it is reasonable to say that data is discrete when tied events occur embarrassingly often.

### 8.4.1 Data formats: wide and long

When working with register data, time is often measured in years which makes it necessary and convenient to work with discrete models. A typical data format is the so-called *wide* format, where there is one record (row) per individual, and measurements for many years. We have so far only worked with the *long* format. The data sets created by `survSplit` are in long format; there is one record per individual and age category. The R work horse in switching back and forth between the long and wide formats is the function `reshape`. It may look confusing at first, but if data follow some simple rules, it is quite easy to use `reshape`.

The function `reshape` is typically used with *longitudinal data*, where there are several measurements at different time points for each individual. If the data for one individual is registered within one record (row), we say that data are in wide format, and if there is one record (row) per time (several records per individual), data are in long format. Using wide format, the rule is that time-varying variable names must end in a numeric value indicating at which time the measurement was taken. For instance, if the variable `civ` (civil status) is noted at times 1, 2, and 3, there must be variables named civ.1, civ.2, and civ.3, respectively. It is optional to use any *separator* between the base name (civ) and the time, but it should be one character or empty. The "." is what reshape expects by default, so using that form simplifies coding somewhat.

We start by creating an example data set as an illustration. This is accomplished by starting off with the data set oldmort in eha and "trimming" it.

```
data(oldmort)
om <- oldmort[oldmort$enter == 60, ]
om <- age.window(om, c(60, 70))
om$m.id <- om$f.id <- om$imr.birth <- om$birthplace <- NULL
om$birthdate <- om$ses.50 <- NULL
om1 <- survival::survSplit(om, cut = 61:69, start = "enter",
                           end = "exit", event = "event",
                           episode = "agegrp")
om1$agegrp <- factor(om1$agegrp, labels = 60:69)
om1 <- om1[order(om1$id, om1$enter), ]
rownames(om1) <- 1:NROW(om1)
om1$id <- as.numeric(as.factor(om1$id))
head(om1)
```

|   | id | sex | civ | region | enter | exit | event | agegrp |
|---|----|-----|-----|--------|-------|------|-------|--------|
| 1 | 1 | male | widow | rural | 60 | 61.000 | 0 | 60 |
| 2 | 1 | male | widow | rural | 61 | 62.000 | 0 | 61 |
| 3 | 1 | male | widow | rural | 62 | 63.000 | 0 | 62 |
| 4 | 1 | male | widow | rural | 63 | 63.413 | 0 | 63 |
| 5 | 2 | female | widow | industry | 60 | 61.000 | 0 | 60 |
| 6 | 2 | female | widow | industry | 61 | 62.000 | 0 | 61 |

This is the long format, each individual has as many records as "presence ages". For instance, person No. 1 has four records, for the ages 60–63. The maximum possible No. of records for one individual is 10. We can check the distribution of No. of records per person by using the function tapply:

```
recs <- tapply(om1$id, om1$id, length)
table(recs)
```

```
recs
  1   2   3   4   5   6   7   8   9  10
400 397 351 315 307 250 192 208 146 657
```

It is easier to get to grips with the distribution with a graph, in this case a *barplot*, see Figure 8.14.

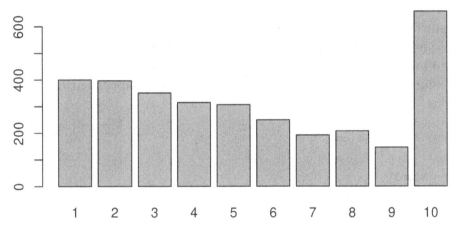

**FIGURE 8.14** Barplot of the number of records per person.

Now, let us turn om1 into a data frame in *wide* format. This is done with the function reshape. First we remove the redundant variables enter and exit.

```
om1$exit <- om1$enter <- NULL
om2 <- reshape(om1, v.names = c("event", "civ", "region"),
               idvar = "id", direction = "wide",
               timevar = "agegrp")
names(om2)
```

```
 [1] "id"        "sex"       "event.60" "civ.60"   "region.60"
 [6] "event.61" "civ.61"    "region.61" "event.62" "civ.62"
[11] "region.62" "event.63" "civ.63"    "region.63" "event.64"
[16] "civ.64"    "region.64" "event.65" "civ.65"   "region.65"
[21] "event.66" "civ.66"    "region.66" "event.67" "civ.67"
[26] "region.67" "event.68" "civ.68"    "region.68" "event.69"
[31] "civ.69"    "region.69"
```

Here there are two time-fixed variables, id and sex, and three time-varying variables, event, civ, and region. The time-varying variables have suffix of the type .xx, where xx varies from 60 to 69.

This is how data in wide format usually show up; the suffix may start with something else than ., but it must be a single character, or nothing. The real problem is how to switch from wide format to long, because our survival analysis tools want it that way. The solution is to use reshape again, but with other specifications.

```
om3 <- reshape(om2, direction = "long", idvar = "id",
               varying = 3:32)
head(om3)
```

```
      id    sex time event      civ   region
1.60   1   male   60     0    widow    rural
2.60   2 female   60     0    widow industry
3.60   3   male   60     0  married     town
4.60   4   male   60     0  married     town
5.60   5 female   60     0  married     town
6.60   6 female   60     0  married     town
```

There is a new variable time created, which goes from 60 to 69, one step for each of the ages. We would like to have the file sorted primarily by id and secondary by time.

```
om3 <- om3[order(om3$id, om3$time), ]
om3[1:11, ]
```

|        | id | sex    | time | event | civ         | region      |
|--------|----|--------|------|-------|-------------|-------------|
| 1.60   | 1  | male   | 60   | 0     | widow       | rural       |
| 1.61   | 1  | male   | 61   | 0     | widow       | rural       |
| 1.62   | 1  | male   | 62   | 0     | widow       | rural       |
| 1.63   | 1  | male   | 63   | 0     | widow       | rural       |
| 1.64   | 1  | male   | 64   | NA    | <NA>        | <NA>        |
| 1.65   | 1  | male   | 65   | NA    | <NA>        | <NA>        |
| 1.66   | 1  | male   | 66   | NA    | <NA>        | <NA>        |
| 1.67   | 1  | male   | 67   | NA    | <NA>        | <NA>        |
| 1.68   | 1  | male   | 68   | NA    | <NA>        | <NA>        |
| 1.69   | 1  | male   | 69   | NA    | <NA>        | <NA>        |
| 2.60   | 2  | female | 60   | 0     | widow       | industry    |

Note that all individuals got 10 records here, even those who only are observed for fewer years. Individual No. 1 is only observed for the ages 60–63, and the next six records are redundant; they will not be used in an analysis if kept, so it is from a practical point of view a good idea to remove them.

```
NROW(om3)
```

```
[1] 32230
```

```
om3 <- om3[!is.na(om3$event), ]
NROW(om3)
```

```
[1] 17434
```

The data frame shrunk to almost half of what it was originally. First, let us summarize data.

```
summary(om3)
```

```
        id              sex              time              event
 Min.   :   1    male  : 7045    Min.   :60.00    Min.   :0.00000
 1st Qu.: 655    female:10389    1st Qu.:61.00    1st Qu.:0.00000
 Median :1267                    Median :63.00    Median :0.00000
 Mean   :1328                    Mean   :63.15    Mean   :0.02587
 3rd Qu.:1948                    3rd Qu.:65.00    3rd Qu.:0.00000
 Max.   :3223                    Max.   :69.00    Max.   :1.00000
         civ                 region
 unmarried: 1565    town    :2485
 married  :11380    industry:5344
 widow    : 4489    rural   :9605
```

The key variables in the discrete time analysis are event and time. For the baseline hazard we need one parameter per value of time, so it is practical to transform the continuous variable time to a factor.

```
om3$time <- as.factor(om3$time)
summary(om3)
```

```
        id              sex            time          event
 Min.   :   1    male  : 7045    60  :3223    Min.   :0.00000
 1st Qu.: 655    female:10389    61  :2823    1st Qu.:0.00000
 Median :1267                    62  :2426    Median :0.00000
 Mean   :1328                    63  :2075    Mean   :0.02587
 3rd Qu.:1948                    64  :1760    3rd Qu.:0.00000
```

```
Max.    :3223               65    :1453   Max.    :1.00000
                         (Other):3674

        civ                  region
unmarried: 1565    town     :2485
married   :11380   industry:5344
widow     : 4489   rural    :9605
```

The summary now produces a frequency table for `time`.

*Note* that we always want our data to be in *long* format before we start the analysis, so the important lesson here was how to go from wide to long. You may find the **tidyr** package (Wickham, 2021) useful if you encounter "untidy" data and want to tidy up.

### 8.4.2  Binomial regression with glm

For a given time point and a given individual, the response is whether an event has occurred or not, that is, it is modeled as a *Bernoulli* outcome, which is a special case of the *binomial* distribution. The discrete time analysis may now be performed in several ways. Most straightforward is to run a *logistic regression* with `event` as response through the basic `glm` function with `family = binomial(link=cloglog)`. The so-called *cloglog* link is used in order to preserve the proportional hazards property in the underlying, real world, continuous time model.

```
fit.glm <- glm(event ~ sex + civ + region + time,
               family = binomial(link = cloglog), data = om3)
summary(fit.glm)
```

```
Call:
```

```
glm(formula = event ~ sex + civ + region + time, family = binomial
(link = cloglog), data = om3)
```

```
Deviance Residuals:
    Min      1Q   Median      3Q      Max
-0.4263  -0.2435  -0.2181  -0.1938   2.9503
```

Coefficients:

|  | Estimate | Std. Error | z value | Pr(>\|z\|) |
|---|---|---|---|---|
| (Intercept) | -3.42133 | 0.21167 | -16.164 | < 2e-16 |
| sexfemale | -0.36332 | 0.09927 | -3.660 | 0.000252 |
| civmarried | -0.33683 | 0.15601 | -2.159 | 0.030847 |
| civwidow | -0.19008 | 0.16810 | -1.131 | 0.258175 |
| regionindustry | 0.10784 | 0.14443 | 0.747 | 0.455264 |
| regionrural | -0.22423 | 0.14034 | -1.598 | 0.110080 |
| time61 | 0.04895 | 0.18505 | 0.265 | 0.791366 |
| time62 | 0.46005 | 0.17400 | 2.644 | 0.008195 |
| time63 | 0.05586 | 0.20200 | 0.277 | 0.782134 |
| time64 | 0.46323 | 0.18883 | 2.453 | 0.014162 |
| time65 | 0.31749 | 0.20850 | 1.523 | 0.127816 |
| time66 | 0.45582 | 0.21225 | 2.148 | 0.031751 |
| time67 | 0.91511 | 0.19551 | 4.681 | 2.86e-06 |
| time68 | 0.68549 | 0.22575 | 3.036 | 0.002394 |
| time69 | 0.60539 | 0.24904 | 2.431 | 0.015064 |

```
(Dispersion parameter for binomial family taken to be 1)

    Null deviance: 4186.8  on 17433  degrees of freedom
Residual deviance: 4125.2  on 17419  degrees of freedom
AIC: 4155.2

Number of Fisher Scoring iterations: 7
```

This output is not so pleasant (we do not want the `time` estimates printed), but we can anyway see that females (as usual) have lower mortality than males, that married are better off than unmarried, and that regional differences maybe are not so large. To get a better understanding of the statistical significance of the findings we run `drop1` on the fit.

```
Single term deletions
```

```
Model:
event ~ sex + civ + region + time
       Df Deviance    AIC    LRT  Pr(>Chi)
<none>           4125.2 4155.2
sex      1       4138.5 4166.5 13.302 0.0002651
civ      2       4130.2 4156.2  4.978 0.0829956
region   2       4135.8 4161.8 10.526 0.0051810
time     9       4161.2 4173.2 36.005 3.956e-05
```

Mildly surprisingly, civil status is not that statistically significant, but region (and the other variables) is. The strong significance of the time variable is of course expected; mortality is expected to increase with increasing age.

### 8.4.3 Survival analysis with coxreg

By some data manipulation we can also use the function coxreg in the package eha for the analysis. For that to succeed we need intervals as responses, and the way of doing that is to add two "fake" variables, exit and enter. The latter must be *slightly* smaller than the former:

```
om3$exit <- as.numeric(as.character(om3$time))
om3$enter <- om3$exit - 0.1
cap = "Old age mortality, discrete time analysis."
fit.ML <- coxreg(Surv(enter, exit, event) ~ sex + civ + region,
              method = "ml", data = om3, coxph = FALSE)
fit.out(fit.ML, caption = cap, label = "mlreg8")
```

The result is shown in Table 8.18 plots of the cumulative hazards and the survival function are easily achieved, see Figures 8.15 and 8.16.

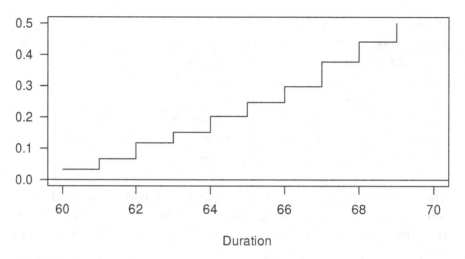

**FIGURE 8.15** The cumulative hazards, from the coxreg fit.

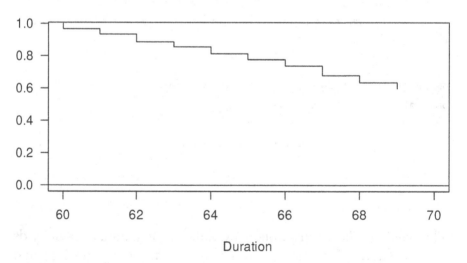

**FIGURE 8.16** The survival function, from the coxreg fit.

**TABLE 8.18** Old age mortality, discrete time analysis.

| Covariate | | Mean | Coef | Rel.Risk | S.E. | L-R p |
|---|---|---|---|---|---|---|
| sex | | | | | | 0.000 |
| | *male* | 0.404 | 0 | 1 | (reference) | |
| | *female* | 0.596 | −0.363 | 0.695 | 0.099 | |
| civ | | | | | | 0.083 |
| | *unmarried* | 0.090 | 0 | 1 | (reference) | |
| | *married* | 0.653 | −0.337 | 0.714 | 0.156 | |
| | *widow* | 0.257 | −0.190 | 0.827 | 0.168 | |
| region | | | | | | 0.005 |
| | *town* | 0.143 | 0 | 1 | (reference) | |
| | *industry* | 0.307 | 0.108 | 1.114 | 0.144 | |
| | *rural* | 0.551 | −0.224 | 0.799 | 0.140 | |
| Events | | 451 | TTR | 1743 | | |
| Max. Log Likelihood | | −2063 | | | | |

Finally, the proportional hazards assumption can be tested in the discrete time framework by creating an interaction between time and the covariates in question. It is possible by using glm.

```
fit2.glm <- glm(event ~ (sex + civ + region) * time,
                family = binomial(link = cloglog),
                data = om3)
drop1(fit2.glm, test = "Chisq")
```

```
Single term deletions

Model:
event ~ (sex + civ + region) * time
            Df Deviance    AIC    LRT Pr(>Chi)
<none>             4077.1 4197.1
sex:time     9     4088.2 4190.2 11.116   0.2679
civ:time    18     4099.5 4183.5 22.425   0.2137
region:time 18     4093.7 4177.7 16.600   0.5508
```

There is no sign of non-proportionality, that is, no interaction with time.

# 9

---

*Multivariate Survival Models*

---

Sometimes survival data come in clusters, and multivariate, or *frailty*, models are appropriate to use.

Ever since the paper by Vaupel et al. (1979), the concept of frailty has spread in even wider circles of the research community. Although their primary purpose was to show various consequences of admitting individual frailties ("individuals age faster than cohorts", due to the selection effect), the effect was that people started to implement their frailty model in Cox regression models.

## 9.1 An Introductory Example

Let us assume that in a follow-up study, the cohort is not homogeneous but instead consists of two equally sized groups with differing hazard rates. Assume further that we have no indication of which group an individual belongs to, and that members of both groups follow an exponential life length distribution:

$$h_1(t) = \lambda_1 \qquad t > 0.$$
$$h_2(t) = \lambda_2$$

This implies that the corresponding survival functions $S_1$ and $S_2$ are

$$S_1(t) = e^{-\lambda_1 t} \qquad t > 0,$$
$$S_2(t) = e^{-\lambda_2 t}$$

DOI: 10.1201/9780429503764-9

and a randomly chosen individual will follow the "population mortality" $S$, which is a *mixture* of the two distributions:

$$S(t) = \frac{1}{2}S_1(t) + \frac{1}{2}S_2(t), \quad t > 0.$$

Let us calculate the hazard function for this mixture. We start by finding the *density function* $f$:

$$f(t) = -\frac{dS(x)}{dx} = \frac{1}{2}\left(\lambda_1 e^{-\lambda_1 t} + \lambda_2 e^{-\lambda_2 t}\right), \quad t > 0.$$

Then, by the definition of $h$ we get

$$h(t) = \frac{f(t)}{S(t)} = w(t)\lambda_1 + (1 - w(t))\lambda_2, \quad t > 0, \qquad (9.1)$$

with

$$w(t) = \frac{e^{-\lambda_1 t}}{e^{-\lambda_1 t} + e^{-\lambda_2 t}}$$

It is easy to see that

$$w(t) \rightarrow \begin{cases} 0, & \lambda_1 > \lambda_2 \\ \frac{1}{2}, & \lambda_1 = \lambda_2 \\ 1, & \lambda_1 < \lambda_2 \end{cases}, \quad \text{as } t \rightarrow \infty,$$

implying that

$$h(t) \rightarrow \min(\lambda_1, \lambda_2), \quad t \rightarrow \infty,$$

see Figure 9.1.

The important point here is that it is *impossible* to tell from data alone whether the population is homogeneous, with all individuals following the same hazard function given by equation (9.1), or if it in fact consists of two groups, each following a constant hazard rate. Therefore, individual frailty models like $h_i(t) = Z_i h(t)$, $i = 1, \ldots, n$, where $Z_i$ is the "frailty" for individual No. $i$, and $Z_1, \ldots, Z_n$ are independent and identically distributed (iid) are less useful.

A heuristic explanation to all this is the dynamics of the problem: We follow a population (cohort) over time, and the *composition*

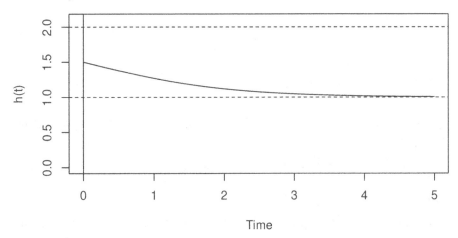

**FIGURE 9.1** Population hazard function (solid line). The dashed lines are the hazard functions of each group, constant at 1 and 2.

of it changes over time. The weaker individuals die first, and the proportion stronger will steadily grow as time goes by.

Another terminology is to distinguish between *individual* and *population* hazards. In Figure 9.1 the solid line is the population hazard, and the dashed lines represent the two kinds of individual hazards present. Of course, in a truly homogeneous population, these two concepts coincide.

## 9.2 Frailty Models

Frailty models in survival analysis correspond to *hierarchical* models in linear or generalized linear models. They are also called *mixed effects models*. A general theory, with emphasis on using **R**, of mixed effects models can be found in Pinheiro and Bates (2000).

### 9.2.1 The simple frailty model

Vaupel et al. (1979) described an individual frailty model,

$$h(t; \mathbf{x}, Z) = h_0(t) Z e^{\beta \mathbf{x}}, \quad t > 0,$$

where $Z$ is assumed to be drawn independently for each individual. Hazard rates for "random survivors" are not proportional, but converging (to each other) if the frailty distribution has finite variance. Thus, the problem may be less pronounced in AFT than in PH regression. However, as indicated in the introductory example, with individual frailty the identification problems are large, and such models are best avoided.

### 9.2.2 The shared frailty model

Frailty models work best when there is a natural grouping of the data, so that observations from the same group are dependent, while two individual survival times from different groups can be regarded as independent. Such a model may be described as

$$h_i(t; \mathbf{x}) = h_{i0}(t) e^{\beta \mathbf{x}}, \quad i = 1, \dots, s; \ t > 0, \tag{9.2}$$

which simply is a stratified Cox regression model. By assuming

$$h_{i0}(t) = Z_i h_0(t), \quad i = 1, \dots s; \ t > 0, \tag{9.3}$$

the traditional multivariate frailty model emerges. Here it is assumed that $Z_1, \dots, Z_s$ are independent and identically distributed (*iid*), usually with a lognormal distribution. From equations (9.2) and (9.3) we get, with $U_i = \log(Z_i)$,

$$h_i(t; \mathbf{x}) = h_0(t) e^{\beta \mathbf{x} + U_i}, \quad i = 1, \dots, s; \ t > 0.$$

In this formulation, $U_1, \dots, U_s$ are *iid* normal with mean zero and unknown variance $\sigma^2$. Another popular choice of distribution for the $Z$:s is the gamma distribution.

In **R**, the package coxme (Therneau, 2020) fits frailty models. We look at the fertility data set in the **R** package eha, see Table 9.1.

**TABLE 9.1** The fertility data set.

| id | parity | age | year | next.ivl | event | prev.ivl | ses | parish |
|----|--------|-----|------|----------|-------|----------|---------|--------|
| 1 | 0 | 24 | 1825 | 0.411 | 1 | NA | farmer | SKL |
| 1 | 1 | 25 | 1826 | 22.348 | 0 | 0.411 | farmer | SKL |
| 2 | 0 | 18 | 1821 | 0.304 | 1 | NA | unknown | SKL |
| 2 | 1 | 19 | 1821 | 1.837 | 1 | 0.304 | unknown | SKL |
| 2 | 2 | 21 | 1823 | 2.546 | 1 | 1.837 | unknown | SKL |

It seems natural to assume that the lengths of birth intervals vary with mother, so we try a frailty model with id (mother's id) as the grouping variable. Also notice that the first interval for each woman is measured from marriage (only married women are included in this data set) to first birth, so we will start by removing them. They are characterized by parity being 0.

```
fe <- fert[fert$parity != 0, ]
library(coxme)
fit <- coxme(Surv(next.ivl, event) ~ age + ses + parity +
             (1 | id), data = fe)
summary(fit, digits = 3)
```

```
Cox mixed-effects model fit by maximum likelihood
  Data: fe
  events, n = 8458, 10312
  Iterations= 15 65
                   NULL Integrated     Fitted
Log-likelihood -71308.31   -69905.17 -68211.87

                 Chisq   df p  AIC   BIC
Integrated loglik  2806    6 0 2794  2752
 Penalized loglik  6193 1212 0 3770 -4764

Model:  Surv(next.ivl, event) ~ age + ses + parity + (1 | id)
Fixed coefficients
```

```
              coef exp(coef) se(coef)      z       p
age          -0.0796    0.923  0.00417 -19.11 0.00000
sesunknown   -0.0779    0.925  0.06067  -1.28 0.20000
sesupper      0.1009    1.106  0.15799   0.64 0.52000
seslower     -0.1703    0.843  0.04967  -3.43 0.00061
parity       -0.1141    0.892  0.01014 -11.26 0.00000


Random effects
 Group Variable  Std Dev Variance
 id     Intercept 0.752    0.565
```

The estimates of the fixed effects have the same interpretation as in ordinary Cox regression. The question is if the results point to the significance of including frailty terms? In the last line of the output we get the estimate of the frailty variance, $\sigma^2 = 0.565$, but no $p$-value for the test of the null hypothesis $H_0 : \sigma = 0$. One explanation to this is that ordinary asymptotic theory does not hold for parameter values at the boundary of the parameter space, the value $\sigma = 0$ is on that boundary.

One way to get a feeling for the impact of the frailty effect is to fit the same model but without frailty, i.e., the term (1 | id), see Table 9.2.

```
fit0 <- coxreg(Surv(next.ivl, event) ~ age + ses + parity,
               data = fe)
cap = "Fertility data, no-frailty model."
lab = "nofrail9"
fit.out(fit0, caption = cap, label = lab)
```

We can compare the two "max log likelihoods", in the frailty model the "Integrated" value –69905, and in the fixed effects case –70391. The difference is so large (486) that we safely can reject the hypothesis that the frailty model is not needed. As an "informal" test, you could take twice that difference and treat it as as a $\chi^2$ statistic with 1 degree of freedom, calculate a $p$-value and take as

**TABLE 9.2** Fertility data, no-frailty model.

| Covariate | | Mean | Coef | Rel.Risk | S.E. | L-R $p$ |
|---|---|---|---|---|---|---|
| age | | 33.088 | −0.080 | 0.923 | 0.003 | 0.000 |
| ses | | | | | | 0.000 |
| | *farmer* | 0.491 | 0 | 1 | (reference) | |
| | *unknown* | 0.186 | −0.085 | 0.919 | 0.030 | |
| | *upper* | 0.018 | 0.147 | 1.158 | 0.083 | |
| | *lower* | 0.305 | −0.083 | 0.920 | 0.026 | |
| parity | | 4.242 | 0.013 | 1.013 | 0.007 | 0.068 |
| Events | | 8458 | TTR | 29806 | | |
| Max. Log Likelihood | | −70391 | | | | |

**TABLE 9.3** The old age mortality data.

| enter | exit | event | birthdate | m.id | sex | civ |
|---|---|---|---|---|---|---|
| 60 | 60.573 | FALSE | 1819.427 | NA | male | widow |
| 60 | 60.371 | FALSE | 1819.629 | NA | female | unmarried |
| 60 | 60.275 | FALSE | 1819.725 | NA | male | married |
| 60 | 60.001 | FALSE | 1819.999 | 782001020 | male | married |
| 60 | 63.766 | FALSE | 1816.234 | NA | female | married |

real $p$-value one half of that (all this because ordinary asymptotic theory does not hold for parameter values on the boundary of the parameter space!). This gives an approximation of the true $p$-value that is not too bad.

As a final example, let us look back at *old age mortality* in the **R** package eha. This example also shows a dangerous situation that is too easy to overlook. It has nothing to do with frailty, but with a problem caused by *missing data*.

Take a look at the variables in oldmort (Table 9.3):

The variable m.id is *mother's id*. Siblings will have the same value on that variable, and we can check whether we find a "sibling effect" in the sense that siblings tend to have a similar risk of dying.

Cox mixed-effects model fit by maximum likelihood

```
Data: oldmort
events, n = 888, 3340 (3155 observations deleted due to missingness)
Iterations= 16 84
                      NULL Integrated      Fitted
Log-likelihood -5702.662    -5693.43 -5646.252


                   Chisq      df         p  AIC      BIC
Integrated loglik  18.46    4.00 1.0012e-03 10.46    -8.69
 Penalized loglik 112.82   49.23 6.7155e-07 14.36 -221.39


Model:  Surv(enter, exit, event) ~ sex + civ + (1 | m.id)
Fixed coefficients
                  coef exp(coef)    se(coef)     z       p
sexfemale  -0.2026325 0.8165783 0.07190559 -2.82 0.00480
civmarried -0.4653648 0.6279060 0.12144548 -3.83 0.00013
civwidow   -0.3305040 0.7185615 0.11964284 -2.76 0.00570


Random effects
 Group Variable Std Dev    Variance
 m.id  Intercept 0.23719204 0.05626007
```

Now, compare with the corresponding fixed effects model in Table 9.4.

**TABLE 9.4** Old age mortality, fixed effects model.

| Covariate | | Mean | Coef | Rel.Risk | S.E. | L-R p |
|---|---|---|---|---|---|---|
| sex | | | | | | 0.000 |
| | *male* | 0.406 | 0 | 1 | (reference) | |
| | *female* | 0.594 | −0.243 | 0.784 | 0.047 | |
| civ | | | | | | 0.000 |
| | *unmarried* | 0.080 | 0 | 1 | (reference) | |
| | *married* | 0.530 | −0.397 | 0.672 | 0.081 | |
| | *widow* | 0.390 | −0.261 | 0.770 | 0.079 | |
| Events | | 1971 | TTR | 37824 | | |
| Max. Log Likelihood | | −13558 | | | | |

Note that we now got a very much smaller value of the maximized log likelihood, −13558 compared to −5693! Something is wrong, and the big problem is that the two analyzes were performed on different data sets. How is that possible, we used oldmort on both occasions? The variable m.id has a lot of missing values, almost 50 per cent are missing (NA), and the standard treatment of NA:s in R is to simply remove each record that contains an NA on any of the variables in the analysis. So, in the first case, the frailty model, a lot of records are removed before analysis, but not in the second. To be able to compare the models we must remove all records with m.id = NA from the second analysis, see Table 9.5.

```
olm <- oldmort[!is.na(oldmort$m.id), ]
fit0 <- coxreg(Surv(enter, exit, event) ~ sex + civ,
               data = olm)
```

**TABLE 9.5** Old age mortality for persons with known mother.

| Covariate | | Mean | Coef | Rel.Risk | S.E. | L-R p |
|---|---|---|---|---|---|---|
| sex | | | | | | 0.006 |
| | male | 0.418 | 0 | 1 | (reference) | |
| | female | 0.582 | −0.196 | 0.822 | 0.070 | |
| civ | | | | | | 0.001 |
| | unmarried | 0.076 | 0 | 1 | (reference) | |
| | married | 0.555 | −0.443 | 0.642 | 0.118 | |
| | widow | 0.369 | −0.310 | 0.733 | 0.116 | |
| Events | | 888 | TTR | 19855 | | |
| Max. Log Likelihood | | −5694 | | | | |

This is another story! We now got very similar values of the maximized log likelihoods, −5693.8 compared to −5693.4! The conclusion is that in this case, there is no frailty effect whatsoever.

One lesson to learn from this example is that you have to be very cautious when a data set contains missing values. Some functions, like drop1, give a warning when a situation like this is detected, but especially when comparisons are made in more than one step, it is too easy to forget the dangers.

Also note the warning that is printed in the results of coxme: *3163 observations deleted due to missingness.* This is a warning that should be taken seriously. □

### 9.2.3  Parametric frailty models

It is possible utilize the connection between Poisson regression and the piecewise constant proportional hazards model discussed in Chapters 7 and 8 to fit parametric frailty models. We look at the fertility data again. The standard analysis without frailty effects is shown in Table 9.6.

```
fit0 <- pchreg(Surv(next.ivl, event) ~ parity + ses,
               cuts = 0:13, data = fe)
```

**TABLE 9.6** Fertility analysis, fixed effects.

| Covariate | | Mean | Coef | Rel.Risk | S.E. | L-R p |
|---|---|---|---|---|---|---|
| parity | | 4.242 | −0.131 | 0.878 | 0.005 | 0.000 |
| ses | | | | | | 0.000 |
| | *farmer* | 0.491 | 0 | 1 | (reference) | |
| | *unknown* | 0.186 | 0.095 | 1.100 | 0.029 | |
| | *upper* | 0.018 | 0.087 | 1.090 | 0.083 | |
| | *lower* | 0.305 | −0.151 | 0.860 | 0.026 | |
| Events | | 8458 | TTR | 29806 | | |
| Max. Log Likelihood | | −14068 | | | | |
| Restricted mean survival: | | 2.3 | in (0, 13] | | | |

Note the use of the function survival::survSplit (comment # 1 in the code), the transformation to factor for the slices of time (years) created by survSplit, and the creation of an offset (# 3). This is described in detail in Chapter 8.

For testing the presence of a random effect over women, an alternative is to use the function glmmML in the package with the same name.

```
library(glmmML)
fe13 <- survSplit(fe, end = "next.ivl", event = "event",
                  cut = 1:13, episode = "years",
                  start = "start") # 1
fe13$years <- as.factor(fe13$years) # 2
fe13$offs <- log(fe13$next.ivl - fe13$start) # 3
fit1 <- glmmML(event ~ parity + ses + years + offset(offs),
               cluster = id, family = poisson, method = "ghq",
               data = fe13, n.points = 9)
out <- with(fit1, cbind(coefficients, coef.sd))
colnames(out) <- c("Coef", "se(Coef)")
round(out[1:5, ], 3)
```

|             | Coef   | se(Coef) |
|-------------|--------|----------|
| (Intercept) | -3.729 | 0.090    |
| parity      | -0.274 | 0.006    |
| sesunknown  | 0.086  | 0.065    |
| sesupper    | -0.033 | 0.171    |
| seslower    | -0.262 | 0.054    |

The estimated scale parameter in the mixing distribution is 0.851 with standard error 0.024, so the conclusion is very much the same as with the nonparametric (coxme) approach: The clustering effect of *mother* is very strong and must be taken into account in the analysis of birth intervals. The nonparametric approach is easier to use and recommended, but see Section 9.3 for an alternative.

## 9.3   Stratification

A simple way to eliminate the effect of clustering is to *stratify* on the clusters. In the birth intervals example, it would mean that intervals are only compared to other birth intervals from the same mother. The drawback with a stratified analysis is that it is not possible to estimate the effect of covariates that are constant within clusters. In the birth intervals case, it is probable that ses, socio-economic status, would vary little within families. On the other hand, the effect of *birth order* or *mother's age* would be suitable to analyze in a stratified setting, see Table 9.7.

**TABLE 9.7** Fertility data, stratified Cox regression.

| Covariate | Mean | Coef | Rel.Risk | S.E. | L-R p |
|---|---|---|---|---|---|
| parity | 4.242 | −0.329 | 0.720 | 0.008 | 0.000 |
| prev.ivl | 2.423 | −0.054 | 0.948 | 0.016 | 0.001 |
| Events | 8458 | TTR | 29806 | | |
| Max. Log Likelihood | −9558 | | | | |

Contrast this result with an unstratified analysis in Table 9.8.

**TABLE 9.8** Fertility data, unstratified Cox regression.

| Covariate | Mean | Coef | Rel.Risk | S.E. | L-R *p* |
|---|---|---|---|---|---|
| parity | 4.242 | −0.084 | 0.920 | 0.005 | 0.000 |
| prev.ivl | 2.423 | −0.319 | 0.727 | 0.011 | 0.000 |
| Events | 8458 | TTR | 29806 | | |
| Max. Log Likelihood | −70391 | | | | |

Note how the effect of parity is diminished when aggregating comparison over all women, while the effect of the length of the previous interval is enlarged. Try to explain why this result is expected!

Under certain circumstances it is actually possible to estimate an effect of a covariate that is constant within strata, but only if it is interacted with a covariate that is not constant within strata. See the example about maternal and infant mortality in Chapter 10.

# 10

## *Causality and Matching*

Causality is a concept that statisticians and statistical science traditionally shy away from. Recently, however, many successful attempts have been made to include the concept of causality in the statistical theory and vocabulary. A good review of the topic, from a modern, event history analysis point of view, is given in Aalen et al. (2008), Chapter 9. Some of their interesting ideas are presented here.

The traditional standpoint among statisticians was that "we deal with association and correlation, not causality", see Pearl (2000) for a discussion. An exception was the clinical trial, and other situations, where *randomization* could be used as a tool. However, during the last couple of decades, there has been an increasing interest in the possibility to make causal statements even without randomization, that is, in *observational* studies (Rubin, 1974; Robins, 1986).

*Matching* is statistical technique, which has an old history without an explicit connection to causality. However, as we will see, matching is a very important tool in the modern treatment of causality.

In the first edition of this book (Broström, 2012), I wrote

"Unfortunately, the models for event history analysis presented here are not implemented in **R** or, to my knowledge, in any other publicly available software. One exception is *matched data analysis*, which, except the matching itself, can be performed with ordinary stratified Cox regression".

Since then things have changed.

## 10.1   Philosophical Aspects on Causality

The concept of causality has a long history in philosophy, see for instance Aalen et al. (2008) for a concise review. A fundamental question, with a possibly unexpected answer, is "Does everything that happens have a cause?". According to Zeilinger (2005), the answer is "No".

> "The discovery that individual events are irreducibly random is probably one of the most significant findings of the twentieth century. Before this, one could find comfort in the assumption that random events only seem random because of our ignorance ... But for the individual event in quantum physics, not only do we not know the cause, there is no cause."
>
> — Zeilinger (2005)

Of course, this statement must not necessarily be taken literally, but it indicates that nothing is to be taken as granted.

## 10.2   Causal Inference

According to Aalen et al. (2008), there are three major schools in statistical causality, (i) graphical models, (ii) predictive causality, and (iii) counterfactual causality. They also introduce a new concept, *dynamic path analysis*, which can be seen as a merging of (i) and (ii), with the addition that *time* is explicitly entering the models.

## 10.2.1 Graphical models

Graphical models have a long history, emanating from Wright (1921) who introduced *path analysis*. The idea was to show by writing diagrams how variables influence one another. Graphical models has had a revival during the last decades with very active research, see Pearl (2000) and Lauritzen (1996). A major drawback, for event history analysis purposes, is, according to Aalen et al. (2008), that *time* is not explicitly taken into account. Their idea is that causality evolves in time, that is, a *cause* must precede an *effect*.

## 10.2.2 Predictive causality

The concept of predictive causality is based on *stochastic processes*, and that a cause must *precede* an effect in time. This may seem obvious, but very often you do not see it clearly stated. This leads sometimes to confusion, for instance to questions like "What is the cause, and what is the effect?".

One early example is *Granger causality* (Granger, 1969) in time series analysis. Another early example with more relevance in event history analysis is the concept of *local dependence*. It was introduced by Tore Schweder (Schweder, 1970).

Local dependency is exemplified in Figure 10.1.

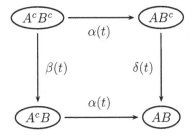

**FIGURE 10.1** Local dependence.

Here $A$ and $B$ are events, and the superscript $(c)$ indicates their complements, i.e., they have not (yet) occurred if superscripted.

This model is used in the matched data example concerning infant and maternal mortality in a 19th century environment later in this chapter. There $A$ stands for *mother dead* and $B$ means *infant dead*. The mother and her new-born (alive) infant is followed from the birth to the death of the infant (but at most a one-year follow-up). During this follow-up both mother and infant are observed and the eventual death of the mother is reported. The question is whether mother's death influences the survival chances of the infant (it does!).

In Figure 10.1: If $\beta(t) \neq \delta(t)$, then $B$ is *locally dependent* on $A$, but $A$ is *locally independent* on $B$: The vertical transition intensities are different, which means that the intensity of $B$ happening is influenced by $A$ happening or not. On the other hand, the horizontal transitions are equal, meaning that the intensity of $A$ happening is not influenced by $B$ happening or not. In our example this means that mother's death influences the survival chances of the infant, but mother's survival chances are unaffected by the eventual death of her infant (maybe not probable in the real world).

### 10.2.3 Counterfactuals

In situations, where interest lies in estimating a *treatment effect* (in a wide sense), the idea of *counterfactual outcomes* is an essential ingredient in the causal inference theory advocated by Rubin (1974) and Robins (1986). A good introduction to the field is given by Hernán and Robins (2020).

Suppose we have a sample of individuals, some treated and some not treated, and we wish to estimate a *marginal* (in contrast to *conditional*) treatment effect in the sample at hand. If the sample is the result of randomization, that is, individuals are randomly allocated to treatment or not treatment (placebo), then there are in principle no problems. If, on the other hand, the sample is self-allocated to treatment or placebo (an observational study), then the risk of *confounders* destroying the analysis is overwhelming. A *confounder* is a variable that is correlated both with treatment and effect, eventually causing biased effect estimates.

The theory of counterfactuals tries to solve this dilemma by allowing each individual to be its own control. More precisely, for each individual, two hypothetical outcomes are defined; the outcome if treated and the outcome if not treated. Let us call them $Y_1$ and $Y_0$, respectively. They are counterfactual (counter to fact), because none of them can be observed. However, since an individual cannot be both treated and untreated, in the real data, each individual has exactly one observed outcome $Y$. If the individual was treated, then $Y = Y_0$, otherwise $Y = Y_1$. The individual treatment effect is $Y_1 - Y_0$, but this quantity is not possible to observe, so how to proceed?

The Rubin school fixes balance in the data by *matching*, while the Robins school advocates *inverse probability weighting*. Both these methods are possible to apply to event history research problems (Hernán et al., 2002, 2005), but unfortunately there is few, if any, publicly available **R** packages for performing these kinds of analyzes, partly with the exception of matching, of which an example is given later in this chapter. However, with the programming power of **R**, it is fairly straightforward to write own functions for specific problems. This is however out of the scope of this presentation.

The whole theory based on counterfactuals relies on the assumption that *there are no unmeasured confounders*. Unfortunately, this assumption is completely un-testable, and even worse, it never holds in practice.

---

## 10.3 Aalen's Additive Hazards Model

In certain applications it may be reasonable to assume that risk factors acts additively rather than multiplicatively on hazards. Aalen's additive hazards model (Aalen, 1989, 1993) takes care of that.

For comparison, recall that the *proportional hazards* model may be written

$$h(t \mid \mathbf{x_i}) = h_0(t)r(\beta, \mathbf{x_i(t)}), \quad t > 0,$$

where $r(\beta, \mathbf{x_i})$ is a *relative risk* function. In *Cox regression* we (usually) have: $r(\beta, \mathbf{x_i}) = \exp(\beta^T \mathbf{x_i})$,

The additive hazards model is given by

$$h(t \mid \mathbf{x_i}) = h_0(t) + \beta_1(t)x_{i1}(t) + \cdots + \beta_p(t)x_{ip}(t), \quad t > 0,$$

where $h_0(t)$ is the *baseline hazard function*, and $\beta(t) = (\beta_0(t), \dots, \beta_p(t))$ is a (multivariate) nonparametric *regression function*.

Note that $h(t, \mathbf{x_i})$ may be negative, if some coefficients or variabless are negative. In contrast to the Cox regression model, there is no automatic protection against this.

The function `aareg` in the `survival` package fits the additive model.

```
Call:
survival::aareg(formula = Surv(enter - 60, exit - 60, event) ~
    sex, data = oldmort)


 n= 6495
    1805 out of 1806 unique event times used

                slope      coef se(coef)     z         p
Intercept     0.1400  0.000753 2.58e-05 29.20 9.82e-188
sexfemale    -0.0281 -0.000136 3.26e-05 -4.16   3.12e-05

Chisq=17.34 on 1 df, p=3.12e-05; test weights=aalen
```

Obviously `sex` is an important variable, females have lower mortality than men. Plots of the time-varying intercept and regression coefficient are given by

```
oldpar <- par(mfrow = c(1, 2))
plot(fit)
par(oldpar)
```

See Figure 10.2, where 95% confidence limits are added around the fitted time-varying coefficients. Also note the use of the function par; the first call sets the plotting area to "one row and two columns" and saves the old par setting in oldpar. Then the plotting area is restored to what it was earlier. It is a good habit to always clean up for the next plotting enterprise.

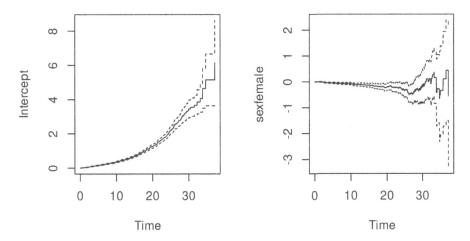

**FIGURE 10.2** Cumulative intercept (left) and cumulative regression coefficient (right).

## 10.4 Dynamic Path Analysis

The term *dynamic path analysis* was coined by Odd Aalen and coworkers (Aalen et al., 2008). It is an extension, by explicitly introducing time, of *path analysis* described by Wright (1921).

The inclusion of time in the model implies that there are one path analysis at each time point, see Figure 10.3.

$X_2$ is an intermediate covariate, while $X_1$ is measured at baseline ($t = 0$). $dN(t)$ is the number of events at $t$.

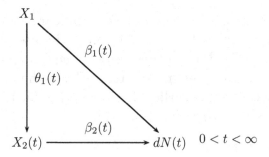

**FIGURE 10.3** Dynamic path analysis.

The *structural equations*

$$dN(t) = (\beta_0(t) + \beta_1(t)X_1 + \beta_2(t)X_2(t))dt + dM(t)$$
$$X_2(t) = \theta_0(t) + \theta_1(t)X_1 + \epsilon(t)$$

are estimated by ordinary least squares (linear regression) at each $t$ with $dN(t) > 0$.

Then the second equation is inserted in the first:

$$dN(t) = \{\beta_0(t) + \beta_2(t)\theta_0(t) + (\beta_1(t) + \theta_1(t)\beta_2(t))X_1$$
$$+ \beta_2(t)\epsilon(t)\}dt + dM(t)$$

and the *total treatment effect* is $(\beta_1(t) + \theta_1(t)\beta_2(t))dt$, so it can be split into two parts according to

$$\text{total effect} = \text{direct effect} + \text{indirect effect}$$
$$= \beta_1(t)dt + \theta_1(t)\beta_2(t)dt$$

Some book-keeping is necessary in order to write an **R** function for this dynamic model. Of great help is the function risksets in eha; it keeps track of the composition of the riskset over time, which is exactly what is needed for implementing the dynamic path analysis. For application where dynamic path analysis is used, see Broström and Bengtsson (2011).

**Estimation:** Standard *least squares* at each $t$, $0 < t < \infty$ (in principle).

## 10.5 Matching

Matching is a way of eliminating the effect of *confounders* and therefore important to discuss in connection with causality. One reason for matching is the wish to follow the causal inference paradigm with counterfactuals. Another reason is that the exposure we want to study is very rare in the population. It will then be inefficient to take a simple random sample from the population; in order to get enough cases in the sample, the sample size needs to be very large. On the other hand, today's register based research tends to analyze whole populations, and the limitations in terms of computer memory and process power are more or less gone. Nevertheless, small properly matched data sets may contain more information about the specific question at hand than whole register!

When creating a matched data set, you have to decide how many controls you want per case. In the true counterfactual paradigm in "causal inference", it is common practice to choose one control per case, to "fill in" the unobservable in the pair of counterfactuals. We first look at the case with matched pairs, then a case study with two controls per case.

### 10.5.1 Paired data

Certain kinds of observations come naturally in pairs. The most obvious situation is data from twin studies. Many countries keep registers of twins, and the obvious advantage with twin data is that it is possible to control for genetic variation; monozygotic twins have identical genes. Humans have pairs of many body parts; arms, legs, eyes, ears, etc. This can be utilized in medical studies concerning comparison of treatments, the two in a pair simply get one treatment each.

In a survival analysis context, pairs are followed over time and it is noted who first experience the event of interest. In each pair, one is treated and the other is a control, but otherwise they are considered more or less identical. Right censoring may result in that it is impossible to decide who in the pair experienced the event first. Such pairs are simply discarded in the analysis.

The model is

$$h_i(t; x) = h_{0i}(t)e^{\beta x}, \qquad (10.1)$$

where $x$ is treatment (0–1) and $i$ is pair No. Note that each pair has its own baseline hazard function, which means that the proportional hazards assumption is only required within pairs. This is a special form of stratified analysis in that each stratum only contains two individuals, a case and its control. If we denote by $T_{i1}$ the life length of the case in pair No. i and $T_{i0}$ the life length of the corresponding control, we get

$$P(T_{1i} < T_{0i}) = \frac{e^{\beta}}{1 + e^{\beta}}, \quad i = 1, \dots, n, \qquad (10.2)$$

and the result of the analysis is simply a study of binomial data; how many times did the case die before the control? We can estimate this probability, and also test the null hypothesis that it is equal to one half, which corresponds to $\beta = 0$ in equation (10.2).

## 10.5.2   More than one control

The situation with more than one control per case is as simple as the paired data case to handle. Simply stratify, with one case and its controls per stratum. It is also possible to have a varying number of controls per case.

As an example where two controls per case were used, let us see how a study of maternal death and its effect on infant mortality was performed.

This is a study on historical data from northern Sweden, 1820–1895 (Broström, 1987). The simple question asked was: How much does the death risk increase for an infant that loses her mother? More precisely, by a maternal death we mean that a mother dies within one year after the birth of her child. In this particular study, only

**TABLE 10.1** Infant and maternal mortality data.

| stratum | enter | exit | event | mother | age | sex | civst | ses | year |
|---|---|---|---|---|---|---|---|---|---|
| 1 | 55 | 365 | 0 | dead | 26 | boy | married | farmer | 1877 |
| 1 | 55 | 365 | 0 | alive | 26 | boy | married | farmer | 1870 |
| 1 | 55 | 365 | 0 | alive | 26 | boy | married | farmer | 1882 |
| 2 | 13 | 76 | 1 | dead | 23 | girl | married | other | 1847 |
| 2 | 13 | 365 | 0 | alive | 23 | girl | married | other | 1847 |
| 2 | 13 | 365 | 0 | alive | 23 | girl | married | other | 1848 |

first births were studied. The total number of such births was 5641 and of these 35 resulted in a maternal death (with the infant still alive). Instead of analyzing the full data set, it was decided to use matching. To each case of maternal death, two controls were selected in the following way. At each time an infant changed status from *mother alive* to *mother dead*, two controls are selected without replacement from the subset of the current risk set, where all infants have *mother alive* and not already used as controls and with correct matching characteristics. If a control changes status to case (its mother dies), it is immediately censored as a control and starts as a case with two controls linked to it. However, this situation never occurred in the study. Let us load the data into **R** and look at it, see Table 10.1.

Here we see the two first triplets in the data set, which consists of 35 triplets, or 105 individuals. Infant No. 1 (the first row) is a case, his mother died when he was 55 days old. At that moment, two controls were selected, that is, two boys 55 days of age, and with the same characteristics as the case. The matching was not completely successful; we had to look a few years back and forth in calendar time (covariate year). Note also that in this triplet all infants survived one year of age, so there is no information of risk differences in that triplet. It will be automatically removed in the analysis.

The second triplet, on the other hand, will be used, because the case dies at age 76 days, while both controls survive one year of age. This is information suggesting that cases have higher mortality than controls after the fatal event of a maternal death.

The matched data analysis is performed by stratifying on triplets (the variable stratum), see Table 10.2.

**TABLE 10.2** Infant mortality, stratified Cox regression.

| Covariate |  | Mean | Coef | Rel.Risk | S.E. | L-R p |
|---|---|---|---|---|---|---|
| mother |  |  |  |  |  | 0.000 |
|  | *alive* | 0.763 | 0 | 1 | (reference) | |
|  | *dead* | 0.237 | 2.605 | 13.534 | 0.757 | |
| Events |  | 21 | TTR | 21616 |  | |
| Max. Log Likelihood |  | −10.8 |  |  |  | |

The result is that mother's death increases the death risk of the infant wit a factor 13.5! It statistically very significant, but the number of infant deaths (21) is very small, so the *p*-value may be unreliable.

In a stratified analysis it is normally not possible to include covariates that are constant within strata. However, here it is possible to estimate the *interaction* between a stratum-constant covariate and exposure, mother's death. However, it is important *not* to include the main effect corresponding to the stratum-constant covariate! This is a rare exception to the rule that when an interaction term is present, the corresponding main effects must be included in the model.

We investigate whether the effect of mother's death is modified by her age by first calculating an interaction term (mage):

```
infants$mage <- ifelse(infants$mother == "dead", infants$age, 0)
```

Note the use of the function ifelse: It takes three arguments, the first is a logical expression resulting in a *logical* vector, with values TRUE and FALSE. Note that in this example, the length is the same as the length of mother (in infants). For each component that is TRUE, the result is given by the second argument, and for each component that is FALSE the value is given by the third argument.

Including the created covariate in the analysis gives the result in Table 10.3.

**TABLE 10.3** Infant mortality and mother's age, stratified Cox regression.

| Covariate | | Mean | Coef | Rel.Risk | S.E. | L-R $p$ |
|---|---|---|---|---|---|---|
| mother | | | | | | 0.562 |
| | *alive* | 0.763 | 0 | 1 | (reference) | |
| | *dead* | 0.237 | 2.916 | 18.467 | 4.860 | |
| mage | | 6.684 | −0.012 | 0.988 | 0.188 | 0.949 |
| Events | | 21 | TTR | 21616 | | |
| Max. Log Likelihood | | −10.8 | | | | |

What happened here? The effect of mother's age is even larger than in the case without mage, but the statistical significance is gone altogether. There are two problems with one solution here. First, due to the inclusion of the interaction, the (main) effect of mother's death is now measured at mother's age equal to 0 (zero!), which of course is completely nonsensical, and second, the construction makes the two covariates strongly correlated: When mother is alive, mage is zero, and when mother is dead, mage takes a rather large value. This is a case of *collinearity*, however not very severe case, because it is very easy to get rid of.

The solution? *Center* mother's age (age)! Recreate the variables and run the analysis again:

```
infants$age <- infants$age - mean(infants$age)
infants$mage <- ifelse(infants$mother == "dead", infants$age, 0)
```

The result is in Table 10.4.

The two fits are equivalent (look at the Max. log. likelihoods), but with different parameterizations. The second, with a centered mother's age, is preferred, because the two covariates are close to uncorrelated there.

**TABLE 10.4** Infant mortality and mother's age, stratified Cox regression, centered mother's age.

| Covariate | | Mean | Coef | Rel.Risk | S.E. | L-R $p$ |
|---|---|---|---|---|---|---|
| mother | | | | | | 0.000 |
| | *alive* | 0.763 | 0 | 1 | (reference) | |
| | *dead* | 0.237 | 2.586 | 13.274 | 0.809 | |
| mage | | 0.280 | −0.012 | 0.988 | 0.188 | 0.949 |
| Events | | 21 | TTR | 21616 | | |
| Max. Log Likelihood | | −10.8 | | | | |

A subject-matter conclusion in any case is that mother's age does not affect the effect of her death on the survival chances of her infant.                                                                    □

**A general advice:** Always center continuous covariates before the analysis! This is especially important in the presence of models with interaction terms, and the advice is valid for all kinds of regression analyzes, not only Cox regression. Common practice is to subtract the mean value, but it is usually a better idea to use a *fixed* value, a value that remains fixed as subsets of the original data set are analyzed.

## 10.6   Conclusion

Finally some thoughts about causality, and the techniques in use for "causal inference". As mentioned above, in order to claim causality, one must show (or assume) that there are "no unmeasured confounders". Unfortunately, this is impossible to prove or show from data alone, but even worse is the fact that in practice, at least in demographic and epidemiological applications, there are *always* unmeasured confounders present. However, with this in mind, note that

- Causal *thinking* is important,
- Counterfactual reasoning and marginal models yield little insight into "how it works", but it is a way of reasoning around research problems that helps sorting out thoughts.
  - Joint modeling is the alternative.
- Creation of *pseudo-populations* through weighting and matching may limit the understanding of how things really work.
  - Analyze the process as it presents itself, so that it is easier to generalize findings.

Read more about this in Aalen et al. (2008).

# 11

## Competing Risks Models

Classical competing risks models will be discussed, as well as their modern interpretations. The basic problem is that we want to consider more than one type of event, but where exactly one will occur. For instance, consider death in different causes.

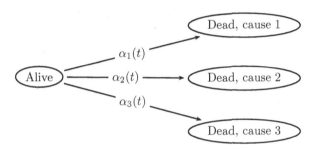

**FIGURE 11.1** Competing risks: Causes of death.

The first problem is: How can the intensities

$$(\alpha_1(t), \alpha_2(t), \alpha_3(t)), \quad t > 0$$

be nonparametrically estimated? It turns out that this is quite simple. The trick is to take one cause at a time and estimate its intensity as if the rest of the causes (events) are censorings. The real problem starts when we want to have these intensities turned into probabilities.

DOI: 10.1201/9780429503764-11

## 11.1   Some Mathematics

First we need some strict definitions, so that the correct questions
are asked, and the correct answers are obtained.

The *cause-specific* cumulative hazard functions are

$$\Gamma_k(t) = \int_0^t \alpha_k(s)ds, \quad t > 0, \quad k = 1, 2, 3.$$

The *total mortality* is

$$\lambda(t) = \sum_{k=1}^{3} \alpha_k(t) \quad t > 0$$

$$\Lambda(t) = \sum_{k=1}^{3} \Gamma_k(t), \quad t > 0$$

and *total survival* is

$$S(t) = \exp\{-\Lambda(t)\}$$

So far, this is not controversial. But asking for a "cause-specific
survivor function" is.

## 11.2   Estimation

The quantities $\Gamma_k$, $k = 1, 2, 3$, $\Lambda$, and $S$ can be estimated in
the usual way. $\Gamma_1$, the *cumulative hazard function for cause 1* is
estimated by regarding all other causes (2 and 3) as censorings. The
total survivor function $S$ is estimated by the method of Kaplan-
Meier, regarding all causes as the same cause (just death).

Is it meaningful to estimate (calculate) $S_k(t) = \exp\{-\Gamma_k(t)\}$, $k =$
$1, 2, 3$? The answer is "No". The reason is that it is difficult to define
what these probabilities mean in the presence of other causes of
death. For instance, what would happen if one cause was eradicated?

## 11.3 Meaningful Probabilities

It *is* meaningful to estimate (and calculate)

$$P_k(t) = \int_0^t S(s)\alpha_k(s)ds, \quad t > 0, \quad k = 1, 2, 3,$$

the probability to die from cause $k$ before time $t$. Note that

$$S(t) + P_1(t) + P_2(t) + P_3(t) = 1 \text{ for all } t > 0.$$

Now, *estimation* is straightforward with the following estimators:

$$\hat{P}_k(t) = \sum_{i:t_i \le t} \hat{S}(t_i-)\frac{d_i^{(k)}}{n_i}$$

See the function `comp` at the end of this chapter for how to write code for this estimation.

## 11.4 Regression

It is also possible to include covariates in the estimating procedure.

$$P_k(t) = 1 - \exp\left\{-\Gamma_k(t)\exp\left(X\beta^{(k)}\right)\right\}, \quad t > 0, \quad k = 1, 2, 3.$$

These equations are estimated separately. In **R**, this can be done with the package `cmprsk` (Fine and Gray, 1999; Gray, 2020).

People born in Skellefteå in northern Sweden around the year 1800 are followed over time from 15 years of age to out-migration or death. Right censoring occurs for those who are still present on January 1, 1870. At exit the cause is noted, see Figure 11.2.

The first rows of the data are shown in Table 11.1.

The variable `event` is coded as shown in Table 11.2.

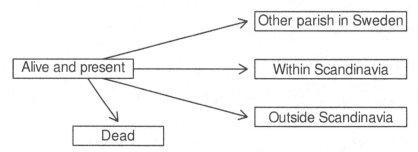

**FIGURE 11.2** Mortality and emigration, Skellefteå, 1800–1870.

**TABLE 11.1** Emigration and mortality data, Skellefteå 1816–1935.

| id | sex | birthdate | enter | exit | event | illeg | soc |
|---|---|---|---|---|---|---|---|
| 9 | male | 1853.391 | 15 | 50 | 0 | FALSE | worker |
| 18 | male | 1845.232 | 15 | 50 | 0 | FALSE | worker |
| 25 | female | 1823.558 | 15 | 50 | 0 | FALSE | worker |
| 28 | male | 1843.633 | 15 | 50 | 0 | FALSE | worker |
| 55 | female | 1830.184 | 15 | 50 | 0 | FALSE | worker |

The code for the function comp is displayed at the end of this chapter. It is not necessary to understand before reading on. It may be useful as a template if you want to analyze competing risks.

However, a simpler alternative is to use the function survfit in the survival package, see Figure 11.4, where the analysis is performed separately for men and women.

Note that the survival is not explicitly shown in Figure 11.4, and

**TABLE 11.2** Competing risks in Skellefteå.

| Code | Meaning | Frequency |
|---|---|---|
| 0 | Censored | 10543 |
| 1 | Dead | 2102 |
| 2 | Moved to other parish in Sweden | 4097 |
| 3 | Moved to other Scandinavian country | 59 |
| 4 | Moved outside the Scandinavian countries | 464 |

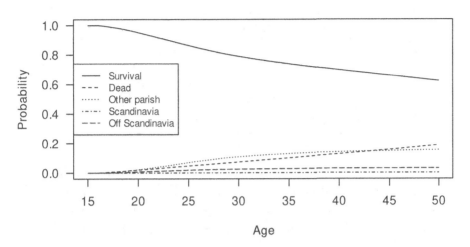

**FIGURE 11.3** Cause-specific exit probabilities; mortality and migration from Skellefteå.

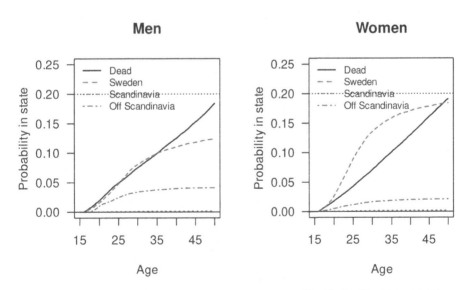

**FIGURE 11.4** Mortality and migration, Skellefteå 1820–1935.

**TABLE 11.3** Output from the function 'crr'.

|             | coef  | exp(coef) | se(coef) | z     | *p*- value |
|-------------|-------|-----------|----------|-------|------------|
| sex:female  | 0.463 | 1.59      | 0.0223   | 20.79 | 0          |
| soc:official| 1.158 | 3.18      | 0.0596   | 22.87 | 0          |
| soc:business| 0.947 | 2.58      | 0.0679   | 13.94 | 0          |
| soc:worker  | 0.566 | 1.76      | 0.0229   | 24.68 | 0          |
| Illegitimate| 0.222 | 1.25      | 0.0487   | 4.56  | 0          |

the reason is that it is *redundant*, the sum of all probabilities at any fixed age in Figure 11.3 is equal to one (an individual must be in one and only one state at each age; note that "survival" really means "survived and still present in Skellefteå", the state all start in at age 15).

And with covariates and cmprsk; the syntax to get it done is a little bit nonstandard. Here *moving to other parish in Sweden* is analyzed (failcode = 2). See Table 11.3.

```
library(cmprsk)
xx <- model.matrix(~ sex + soc + illeg, zz)[, -1]
systim <- system.time(fit <- crr(zz$exit, zz$event, xx, failcode = 2))
summary(fit)
```

The *system time* for this call is 1 hour and 15 minutes. The function crr is *very slow* with large data sets. Therefore I have refrained from calculating likelihood ratio *p*-values, so the ones in Table 11.3 are the *Wald* ones. However, it seems as if all covariates have a high explanatory value, and we note that females move within Sweden more than men, and so do white collar people compared to the rest. Farmers are the most stationary ones, not surprisingly. Maybe more surprising is the fact that people born outside marriage are more prone to migrate, at least within Sweden.

Note that this is *not* Cox regression (but close!). The corresponding Cox regression result is shown in Table 11.4.

```
fit <- coxreg(Surv(enter, exit, event == "Sweden") ~ sex + soc +
              illeg, data = zz)
cap <- "Cox regression, moved to other parish in Sweden."
lab <- "withcox11"
fit.out(fit, caption = cap, label = lab)
```

**TABLE 11.4** Cox regression, moved to other parish in Sweden.

| Covariate | | Mean | Coef | Rel.Risk | S.E. | L-R p |
|---|---|---|---|---|---|---|
| sex | | | | | | 0.000 |
| | *male* | 0.509 | 0 | 1 | (reference) | |
| | *female* | 0.491 | 0.450 | 1.569 | 0.022 | |
| soc | | | | | | 0.000 |
| | *farming* | 0.655 | 0 | 1 | (reference) | |
| | *official* | 0.021 | 1.172 | 3.229 | 0.050 | |
| | *business* | 0.012 | 1.023 | 2.782 | 0.068 | |
| | *worker* | 0.312 | 0.589 | 1.803 | 0.023 | |
| illeg | | | | | | 0.000 |
| | *FALSE* | 0.959 | 0 | 1 | (reference) | |
| | *TRUE* | 0.041 | 0.222 | 1.248 | 0.048 | |
| Events | | 8419 | TTR | 1404716 | | |
| Max. Log Likelihood | | −89750 | | | | |

As can be seen from Table 11.4, the same conclusions can be drawn with an ordinary Cox regression as with the competing risks approach, *in this case*, that is.

What about migrating to countries outside Scandinavia? The same covariates, but stratified on sex, see Table 11.5 and Figure 11.5.

Here illegitimacy does not mean much, but it is the *workers* that dominate migration in this case.

Males end up with a double intensity of migrating, but the females start earlier.

□

**TABLE 11.5** Competing risks by Cox regression and moved outside Scandinavia.

| Covariate | | Mean | Coef | Rel.Risk | S.E. | L-R p |
|---|---|---|---|---|---|---|
| soc | | | | | | NA |
| | farming | 0.655 | 0 | 1 | (reference) | |
| | official | 0.021 | −0.049 | 0.952 | 0.174 | |
| | business | 0.012 | 0.063 | 1.065 | 0.206 | |
| | worker | 0.312 | 0.218 | 1.243 | 0.050 | |
| illeg | | | | | | 0.000 |
| | FALSE | 0.959 | 0 | 1 | (reference) | |
| | TRUE | 0.041 | −0.149 | 0.862 | 0.125 | |
| Events | | 1792 | TTR | 1404716 | | |
| Max. Log Likelihood | | −18147 | | | | |

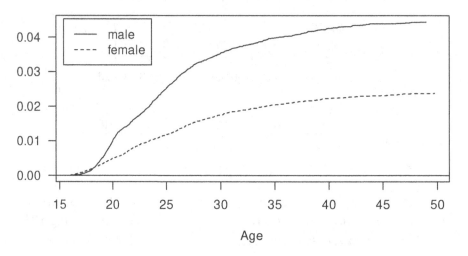

**FIGURE 11.5** Cumulative hazard functions for males and females, migration outside Scandinavia.

## 11.5   R code for Competing Risks

The code that produced Figure 11.3 is shown here.

```
function (enter, exit, event, start.age = 0)
```

```r
{
    par(las = 1)
    library(eha)
    if (is.factor(event)) {
        event <- as.numeric(event) - 1
    }
    n <- max(event)
    rs.tot <- risksets(Surv(enter, exit, event > 0.5))
    haz.tot <- rs.tot$n.events/rs.tot$size
    n.times <- length(haz.tot) + 1
    S <- numeric(n.times)
    S[1] <- 1
    for (i in 2:n.times) S[i] <- S[i - 1] * (1 - haz.tot[i -
        1])
    haz <- matrix(0, nrow = n, ncol = length(haz.tot))
    P <- matrix(0, nrow = n, ncol = length(haz.tot) + 1)
    for (row in 1:n) {
        rs <- risksets(Surv(enter, exit, event == row))
        haz.row <- rs$n.events/rs$size
        tmp <- 0
        cols <- which(rs.tot$risktimes %in% rs$risktimes)
        haz[row, cols] <- haz.row
        P[row, 2:NCOL(P)] <- cumsum(S[1:(n.times - 1)] * haz[row,
            ])
    }
    plot(c(start.age, rs.tot$risktimes), S, ylim = c(0, 1), xlim = c(start.age,
        max(rs.tot$risktimes)), xlab = "Age", type = "s", ylab = "Probability")
    for (i in 1:n) lines(c(start.age, rs.tot$risktimes), P[i,
        ], lty = i + 1, type = "s")
    legend("left", lty = 1:(n + 1), legend = c("Survival", "Dead",
        "Other parish", "Scandinavia", "Off Scandinavia"), cex = 0.8)
    invisible(list(P = P, S = S))
}
<bytecode: 0x5580d9423748>
```

# A

## Basic Statistical Concepts

The statistical concepts that are important for understanding what is going on in this book are gathered here, but briefly treated. The interested reader who wants a deeper understanding of statistical concepts should have no problems in finding suitable text books. There are for instance some texts that teaches statistics and how to use it in **R** (Dalgaard, 2008).

### A.1   Statistical Inference

Statistical inference is the science that help us draw conclusions about real world phenomena by observing and analyzing samples from them. The theory rests on probability theory and the concept of *random* sampling. The statistical analysis never gives absolute truths, but only statements coupled to certain measures of their validity. These measures are almost always *probability statements*.

The crucial concept is that of a *model*, despite the fact that the present trend in statistical inference is toward nonparametric statistics. It is often stated that with today's huge data sets, statistical models are unnecessary, but nothing could be more wrong.

The important idea in a statistical model is the concept of a *parameter*. It is often confused with its estimator from data. For instance, when we talk about *mortality* in a population, it is a hypothetical concept that is different from the ratio between the observed number of deaths and the population size (or any other measure based on data). The latter is an *estimate* (at best) of the

DOI: 10.1201/9780429503764-A

former. The whole idea about statistical inference is to extract information about a *population parameter* from observing data.

### A.1.1 Point estimation

The case in *point estimation* is to find the best guess (in some sense) of a population parameter from data. That is, we try to find the best single value that is closest to the true, but unknown, value of the population parameter.

Of course, a point estimator is useless if it is not connected to some measure of its uncertainty. That takes us to the concept of *interval estimation*.

### A.1.2 Interval estimation

The philosophy behind *interval estimation* is that a guess on a single value of the unknown population parameter is useless without an accompanying measure of the uncertainty of that guess. A *confidence interval* is an interval, which we say covers the true value of the population parameter with a certain prescribed probability (often chosen as 95 per cent).

### A.1.3 Hypothesis testing

We are often interested in a specific value of a parameter, and in regression problems this value is almost always *zero* (0). The reason is that regression parameters measure *effects*, and to test for no effect is then equivalent to testing that the corresponding parameter has value zero.

There is a connection between interval estimation and hypothesis testing: To test the hypothesis that a parameter value is zero can be done through constructing a confidence interval for the parameter. The test rule is then: If the interval does not cover *zero*, reject the hypothesis, otherwise do not.

**TABLE A.1** General table in a log-rank test.

| Group | Deaths | Survivors | Total |
|-------|--------|-----------|-------|
| I     | d1     | n1 – d1   | n1    |
| II    | d2     | n2 – d2   | n2    |
| Total | d      | n – d     | n     |

### A.1.3.1 The log-rank test

The general hypothesis testing theory behind the log-rank test builds on the *hypergeometric distribution*. The calculations under the null hypothesis of no difference in survival chances between the two groups are performed *conditional on both margins*. In Table A.1, if the margins are fixed there is only one degree of freedom left; for a given value of (say) $d_1$, the three values $d_2$, $(n_1 - d_1)$, and $(n_2 - d_2)$ are determined.

Utilizing the fact that, under the null, $d_1$ is hyper-geometrically distributed, results in the following algorithm for calculating a test statistic as follows:

1. Observe $O = d_1$
2. Calculate the expected value $E$ of $O$ (under the null):

$$E = d\frac{n_1}{n}.$$

3. Calculate the variance $V$ of $O$ (under the null):

$$V = \frac{(n - d)dn_1n_2}{n^2(n - 1)}.$$

4. Repeat 1 to 3. for all tables and aggregate according to equation (A.1).

The log rank test statistic $T$ is

$$T = \frac{\sum_{i=1}^{k}(O_i - E_i)}{\sqrt{\sum_{i=1}^{k}V_i}} \tag{A.1}$$

Note carefully that this procedure is *not* equivalent to aggregating all tables of raw data!

Properties of the log rank test;

1. The test statistic $T^2$ is approximately distributed as $\chi^2(1)$.
2. It is available in most statistical software.
3. It can be generalized to comparisons of more than two groups.
4. For $s$ groups, the test statistic is approximately $\chi^2(s-1)$.
5. The test has *high power* against alternatives with *proportional hazards*, but can be weak against non pro por tional alternatives.

## A.2 Presentation of Results

The presentation of the results of fitting a regression model to given data is usually done both graphically and in tabular form. Some general guidelines that must be followed are given here. In order to make the presentation easier to understand, a real data, but fictive research project is followed.

### A.2.1 The project

In a mortality study of the small town of Umeå in 20th century northern Sweden, we are investigating the mortality among persons aged 50 and above. We are especially interested in whether there were any differences in mortality between social branches. For this purpose, the data set ume, with selected variables, is used. A summary of the data is shown in Table A.2.

As you can see, data is *tabulated*, so a piecewise constant hazard model is called for.

**TABLE A.2** Data about adult mortality in Umeå, first half of 20th ccentury.

| sex | cohort | civst | socBranch | age | event | exposure |
|-----|--------|-------|-----------|-----|-------|----------|
| male | [1810,1858) | married | official | 50–53 | 3 | 176.164 |
| female | [1810,1858) | married | official | 50–53 | 3 | 125.279 |
| male | [1858,1875) | married | official | 50–53 | 4 | 419.428 |
| female | [1858,1875) | married | official | 50–53 | 0 | 329.074 |
| male | [1875,1890) | married | official | 50–53 | 12 | 651.818 |

## A.2.2 Tabular presentation

We fit a simple model with only one covariate, but select only *married men*. This can be done with the help of *Poisson regression* and the `glm` function:

```
Call:
glm(formula = event ~ offset(log(exposure)) + age + socBranch,
    family = "poisson", data = umetab, subset = sex == "male" &
        civst == "married")

Deviance Residuals:
    Min      1Q   Median      3Q      Max
-2.7219  -0.9084  -0.1202  0.6842   2.5910

Coefficients:
                Estimate Std. Error z value Pr(>|z|)
(Intercept)     -4.41828    0.09766 -45.240  < 2e-16
age53-56         0.33971    0.11042   3.077  0.00209
age56-59         0.51119    0.10978   4.657 3.21e-06
age59-62         0.83266    0.10635   7.830 4.89e-15
age62-65         1.10338    0.10483  10.526  < 2e-16
age65-68         1.38179    0.10395  13.292  < 2e-16
age68-71         1.80158    0.10196  17.669  < 2e-16
age71-74         1.96600    0.10539  18.655  < 2e-16
age74-77         2.16921    0.11027  19.672  < 2e-16
```

```
age77-80            2.58718    0.11432  22.632  < 2e-16
age80-83            2.66962    0.13539  19.718  < 2e-16
age83-86            3.01993    0.15727  19.202  < 2e-16
age86-89            3.38008    0.19776  17.092  < 2e-16
age89-92            3.26400    0.36312   8.989  < 2e-16
age92-95            3.11514    0.71204   4.375 1.21e-05
socBranchfarming   -0.40721    0.06612  -6.158 7.35e-10
socBranchbusiness   0.03182    0.09226   0.345  0.73017
socBranchworker    -0.32630    0.07125  -4.580 4.65e-06

(Dispersion parameter for poisson family taken to be 1)

    Null deviance: 1752.29  on 192  degrees of freedom
Residual deviance:  240.43  on 175  degrees of freedom
AIC: 969.83

Number of Fisher Scoring iterations: 5
```

Besides the slightly disturbing fact that all age estimates are printed, the presentation of the result for the covariate of interest, socBranch, is less satisfactory from a "survival analysis" point of view: We are generally not interested of looking at the estimated baseline in tabular form (we want a graph). There is a more general critique of the presentation: Too many $p$-values are given, and not the relevant ones either.

In eha, the function tpchreg does the same thing, but with a more satisfactory presentation of the result, see Table A.3.

First, we got rid of the printing of the numerous baseline hazards estimates, and second, the *factor* socBranch is presented in full, with all four levels. And finally, the *relevant p-value*. Because, first, the truth is that *Wald* $p$-values are notoriously unreliably in nonlinear regression, and second, we want $p$-values for *covariates* and *not* for *levels* of covariates.

However, note that the essential findings with the two approaches are identical.

**TABLE A.3** Adult mortality in Umeå, first half of 20th century, PH regression.

| Covariate | | **Mean** | **Coef** | **Rel.Risk** | **S.E.** | **L-R p** |
|---|---|---|---|---|---|---|
| socBranch | | | | | | 0.000 |
| | *official* | 0.109 | 0 | 1 | (reference) | |
| | *farming* | 0.520 | −0.407 | 0.666 | 0.066 | |
| | *business* | 0.072 | 0.032 | 1.032 | 0.092 | |
| | *worker* | 0.299 | −0.326 | 0.722 | 0.071 | |
| Events | | 2266 | TTR | 83599 | | |
| Max. Log Likelihood | | −9686 | | | | |
| Restricted mean survival: | | 20.5 | in (50, 95] | | | |

### A.2.3 Graphics

The baseline hazard function is shown in Figure A.1.

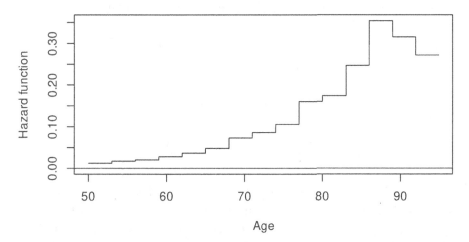

**FIGURE A.1** The hazard function for old age mortality in Umeå, first half of 20th century.

## A.3   Asymptotic Theory

### A.3.1   Partial likelihood

Here is a very brief summary of the asymptotics concerning the partial likelihood. Once defined, it turns out that you may treat it as an ordinary likelihood function Andersen et al. (1993). The setup is as follows.

Let $t_{(1)}, t_{(2)}, \dots, t_{(k)}$ the ordered observed event times and let $R_i = R(t_{(i)})$ be the risk set at $t_{(i)}$, $i = 1, \dots, k$, see equation (2.7). At $t_{(i)}$, *condition* with respect to the composition of $R_i$ and that one event occurred (for tied events, a correction is necessary).

Then the contribution to the partial likelihood from $t_{(i)}$ is

$$L_i(\beta) = P(\text{No. } m_i \text{ dies} \mid \text{one event occur}, R_i)$$
$$= \frac{h_0(t_{(i)}) \exp(\beta \mathbf{x}_{m_i})}{\sum_{\ell \in R_i} h_0(t_{(i)}) \exp(\beta \mathbf{x}_\ell)} = \frac{\exp(\beta \mathbf{x}_{m_i})}{\sum_{\ell \in R_i} \exp(\beta \mathbf{x}_\ell)}$$

and the full partial likelihood is

$$L(\beta) = \prod_{i=1}^{k} L_i(\beta) = \prod_{i=1}^{k} \frac{\exp(\beta \mathbf{x}_{m_i})}{\sum_{\ell \in R_i} \exp(\beta \mathbf{x}_\ell)}$$

This is where the doubt about the partial likelihood comes in; the conditional probabilities multiplied together do not have a proper interpretation as a conditional probability. Nevertheless, it is prudent to proceed as if the expression really is a likelihood function. The log partial likelihood becomes

$$\log\left(L(\beta)\right) = \sum_{i=1}^{k} \left\{ \beta \mathbf{x}_{m_i} - \log\left( \sum_{\ell \in R_i} \exp(\beta \mathbf{x}_\ell) \right) \right\}, \qquad \text{(A.2)}$$

and the components of the *score vector* are

$$\frac{\partial}{\partial \beta_j} \log L(\beta) = \sum_{i=1}^{k} \mathbf{x}_{m_i j} - \sum_{i=1}^{k} \frac{\sum_{\ell \in R_i} x_{\ell j} \exp(\beta \mathbf{x}_\ell)}{\sum_{\ell \in R_i} \exp(\beta \mathbf{x}_\ell)}, \qquad j = 1, \dots, s.$$
$$\text{(A.3)}$$

The *maximum partial likelihood (MPL)* estimator of $\beta$, $\hat{\beta}$, is found by setting (A.3) equal to zero and solve for $\beta$.

For inference, we need to calculate the inverse of minus the *Hessian*, evaluated at $\hat{\beta}$. This gives the estimated *covariance matrix*. The Hessian is the matrix of the second partial derivatives. The expectation of minus the Hessian is called the *information matrix*. The *observed* information matrix is

$$\hat{I}(\hat{\beta})_{j,m} = -\frac{\partial^2 \log L(\beta)}{\partial \beta_j \partial \beta_m}\Big|_{\beta=\hat{\beta}}$$

and asymptotic theory says that

$$\hat{\beta} \sim N(\beta, \hat{I}^{-1}(\hat{\beta}))$$

This is to say that $\hat{\beta}$ is asymptotically unbiased and normally distributed with the given covariance matrix (or the limit of it). Further, $\hat{\beta}$ is a consistent estimator of $\beta$. These results are used for for hypothesis testing, confidence intervals, and variable selection.

Note that these are only *asymptotic* results, i.e., useful in *large to medium sized samples*. In small samples, *bootstrapping* is a possibility. This option is available in the **R** package **eha**.

Here a warning is in order: Tests based on standard errors (*Wald*) tests) may be *highly unreliable*, as in all *non-linear* regression (Hauck and Donner, 1977). A better alternative is the *likelihood ratio test*.

## A.4 Model Selection

In regression models, there is often several competing models for describing data. In general, there are no strict rules for "correct selection". However, for *nested* models, there are some formal guidelines.

## A.4.1  Comparing nested models

The meaning of *nesting* of models is best described by an example.

1.  $M_2 : h(t; (x_1, x_2)) = h_0(t) \exp(\beta_1 x_1 + \beta_2 x_2)$
2.  $M_1 : h(t; (x_1, x_2)) = h_0(t) \exp(\beta_1 x_1)$: $x_2$ has no effect.

Thus, the model $M_1$ is a special case of $M_2$ ($\beta_2 = 0$). We say that $M_1$ is *nested* in $M_2$. Now, *assume* that $M_2$ is *true*. Then, testing the hypothesis $H_0 : M_1$ is true (as well) is the same as testing the hypothesis $H_0; \beta_2 = 0$.

The formal theory for and procedure for performing the likelihood ratio test (LRT) can be summarized as follows:

1.  Maximize $\log L(\beta_1, \beta_2)$ under $M_2$; gives $\log L(\hat{\beta}_1, \hat{\beta}_2)$.

2.  Maximize $\log L(\beta_1, \beta_2)$ under $M_1$, that is, maximize $\log L(\beta_1, 0)$; gives $\log L(\beta_1^*, 0)$.

3.  Calculate the test statistic

$$T = 2( \log L(\hat{\beta}_1, \hat{\beta}_2) - \log L(\beta_1^*, 0))$$

4.  Under $H_0$, $T$ has a $\chi^2$ (chi-square) distribution with $d$ degrees of freedom: $T \sim \chi^2(d)$, where $d$ is the difference in numbers of parameters in the two competing models, in this case $2 - 1 = 1$.

5.  Reject $H_0$ if $T$ is large enough. Exactly how much that is depends on the level of significance; if it is $\alpha$, choose the limit $t_d$ equal to the $100(1 - \alpha)$ percentile of the $\chi^2(d)$ distribution.

This result is a *large sample approximation*.

The *Wald test* is theoretically performed as follows:

1.  Maximize $\log L(\beta_1, \beta_2)$ under $M_2$; this gives $\log L(\hat{\beta}_1, \hat{\beta}_2)$, and $\hat{\beta}_2$, se($\hat{\beta}_2$).

2. Calculate the test statistic

$$T_W = \frac{\hat{\beta}_2}{\text{se}(\hat{\beta}_2)}$$

3. Under $H_0$, $T_W$ has a *standard normal* distribution: $T_W \sim N(0,1)$.

4. Reject $H_0$ if the absolute value of $T_W$ is larger than 1.96 on a significance level of 5%.

This is a *large sample approximation*, with the advantage that it is automatically available in all software. In comparison to the LRT, one model less has to be fitted. This saves time and efforts, unfortunately on the expense of accuracy, because it may occasionally give *nonsensic results*. This phenomenon, as already mentioned, is known as the *Hauck-Donner effect* (Hauck and Donner, 1977). □

## A.4.2 Comparing non-nested models

Non-nested models cannot be compared by a likelihood ratio test, but there are a couple of alternatives that are based on comparing maximized likelihood values modified with consideration of the number of parameters that needs to be estimated. One such alternative is the *Akaike Information Criterion (AIC)*, see de Leeuw (1992).

# B

## Survival Distributions

The survival distributions we discuss here are all available in **R**. The most basic survival distribution is the *Exponential distribution*, not because it is useful in demographic applications (it isn't), but because it is a reference distribution, a special case of other distributions, and a building block in many useful models.

The piecewise constant hazard, Gamma, and the Weibull distributions are all generalizations of the exponential.

### B.1   Relevant Distributions in R

In **R**, there are several families of distributions available. The ones that are relevant in survival analysis are characterized by having positive support. For each family of distributions, four functions are available; a *density* function (name prefix d), a *cumulative distribution* function (name prefix p), a *quantile* function (name prefix q), and a *random number* generator (name prefix r). In the package **eha**, the two functions with prefix h and H are added for some distributions. For instance, for the *Weibull* distribution there are hweibull and Hweibull, where the first is the hazard function and the second is the cumulative hazards function.

We show exactly what these functions are and how to use them. It should be noted that in base **R**, there are no survival functions, hazard functions, or cumulative hazards functions. Some are available in the package **eha** and others. However, they can easily be derived from the given p- and d- functions, as shown in the next subsection B.1.1.

DOI: 10.1201/9780429503764-B

### B.1.1   The Exponential distribution

The exponential distribution is characterized by having a constant hazard rate $\lambda$. It has density function

$$f(t; \lambda) = \lambda e^{-\lambda x}, \quad \lambda > 0, \, t > 0,$$

and survival function

$$S(t; \lambda) = e^{-\lambda x}, \quad \lambda > 0, \, t > 0.$$

The mean is $1/\lambda$ (which also is the *scale* parameter), and the variance is $1/\lambda^2$. The exponential distribution is represented in **R** by the functions pexp, dexp, qexp, and rexp, in order the cumulative distribution function, the density function, the quantile function (which is the inverse to the cumulative distribution function), and the random number generator function.

In Figure B.1 the relevant functions are plotted for an exponential distribution with $\lambda = 1$ (the default value). It is created as follows.

```
library(eha)
x <- seq(0, 4, length = 1000)
oldpar <- par(mfrow = c(2, 2))
plot(x, dexp(x), type = "l",
     main = "Density", ylab = "")
plot(x, pexp(x, lower.tail = FALSE), type = "l",
     main = "Survival", ylab = "")
plot(x, hweibull(x, shape = 1), type = "l",
     main = "Hazard", ylab = "")
plot(x, Hweibull(x, shape = 1), type = "l",
     main = "Cumulative hazards", ylab = "")
par(oldpar)
```

Note that there are no functions hexp or Hexp, maybe because they are too simple, see Figure B.1. The fact that the exponential distribution is special case of the Weibull distribution (shape = 1) is utilized.

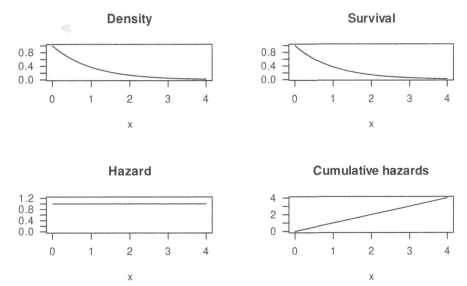

**FIGURE B.1** The exponential distribution with scale parameter 1.

The exponential distribution is characterized by the fact that it *lacks memory*. In other words, items whose life lengths follow an exponential distribution do not age; no matter how old they are, if they are alive they are as good as new. This concept is not useful when it comes to human lives, but the life lengths of electronic components are often modeled by the exponential distribution in reliability theory.

### B.1.2 The piecewise constant hazard distribution

If the exponential distribution is not useful in describing human lives, it may be so for short segments of life. At least it will be a good approximation if the segment is short enough.

This is the idea behind the *piecewise constant hazard* distribution, called pch in **eha**. Its definition involves a partition of the time (age) axis, and one positive constant (the hazard level) corresponding to each interval. Note that the last interval will be open, with infinite length; only a finite number of cutpoints are allowed. The definition of the hazard function $h$ becomes, with the cuts denoted

$\mathbf{t} = (t_1, < \cdots < t_n)$ and the levels denoted $\mathbf{h} = (h_1, \ldots, h_{n+1})$:

$$h(t; \mathbf{t}, \mathbf{h}) = \begin{cases} h_1 & t \le t_1, \\ h_i & t_{i-1} < t \le t_i, \; i = 2, \ldots, n, \\ h_{n+1} & t_n < t. \end{cases} \qquad (\text{B.1})$$

In this definition, the number of levels must be exactly one more than the number of cutpoints. The relevant functions are shown in Figure B.2, created as follows:

```
cuts <- c(1, 2, 3)
n <- length(cuts)
levels <- c(3, 2, 1, 2)
oldpar <- par(mfrow = c(2, 2), las = 1)
plot(x, dpch(x, cuts = cuts, levels = levels),
     type = "l", main = "Density", ylab = "", xlab = "t")
plot(x, ppch(x, cuts = cuts, levels = levels, lower.tail = FALSE),
     type = "l", main = "Survival", ylab = "", xlab = "t")
##plot(x, hpch(x, cuts = cuts, levels = levels),
plot(c(0, cuts[1]), c(levels[1], levels[1]),
     type = "l", main = "Hazard", ylab = "", xlab = "t",
     ylim = c(0, max(levels) + 0.2), xlim = c(0, max(x) + 0.2))
for (i in 1:(n - 1)){
    lines(c(cuts[i], cuts[i+1]), c(levels[i+1], levels[i+1]))
}
lines(c(cuts[n], Inf, c(levels[n+1], levels[n+1])))
plot(x, Hpch(x, cuts = cuts, levels = levels),
     type = "l", main = "Cumulative hazards",
     ylab = "", xlab = "t")



par(oldpar)
```

Note that, despite the fact that the hazard function is *not* continuous, the other functions are. They are not differentiable at the cut points, though.

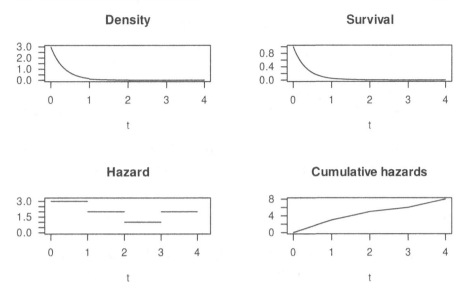

**FIGURE B.2** Piecewise constant hazard distribution.

The piecewise constant hazard distribution is very flexible. It can be made arbitrarily close to any continuous distribution by increasing the number of cutpoints and choosing the levels appropriately. Parametric proportional hazards modeling with the *pch* distribution is a serious competitor to the Cox regression model, especially with large data sets.

### B.1.3 The Weibull distribution

The *Weibull* distribution is a very popular parametric model for survival data, described in detail by Waloddi Weibull (Weibull, 1951), but known earlier. It is one of the so-called extreme-value distributions, and as such very useful in reliability theory. It is becoming popular in demographic applications, but in mortality studies it is wise to avoid it for adult mortality (the hazard grows too slow) and mortality in ages 0–15 years of age (U-shaped hazards, which the Weibull model doesn't allow).

The hazard function of a Weibull distribution is defined by

$$h(t; (p, \lambda)) = \frac{p}{\lambda}\left(\frac{t}{\lambda}\right)^{p-1}, \quad t, p, \lambda > 0.$$

where $p$ is a *shape* parameter and $\lambda$ is a *scale* parameter. When $p = 1$, this reduces to $h(t; (1, \lambda) = 1/\lambda$, which is the exponential distribution with rate $1/\lambda$. Compare to the definition of the exponential distribution and note that there $\lambda$ is the *rate* and here it is a *scale* parameter, which is the inverted value of the rate.

As mentioned above, the Weibull distribution has a long history in reliability theory. Early on simple graphical tests were constructed for judging if a real data set could be adequately described by the Weibull distribution. The so-called *Weibull paper* was invented. Starting with the definition of the *cumulative* hazards function $H$,

$$H(t; (p, \lambda)) = \left(\frac{t}{\lambda}\right)^p,$$

by taking logarithms of both sides we get

$$\ln H(t; (p, \lambda)) = p \ln(t) - p \ln(\lambda)$$

with $x = \ln t$ and $y = \ln H(t)$, this is a straight line. So by plotting the Nelson-Aalen estimator of the cumulative hazards function against log time, it is possible to graphically check the Weibull and exponential assumptions.

Is it reasonable to assume that the survival data in the data set mort can be modeled by the Weibull distribution? We construct the *Weibull plot* with the following code.

```
fit <- coxreg(Surv(enter, exit, event) ~ 1, data = mort)
'plot(fit, fn = "loglog")'
```

The result is shown in Figure B.3.

Except for the early disturbance, linearity is not too far away (remember that the log transform magnifies the picture close to zero). Is the slope close to 1? Probably not, the different scales on the axis makes it hard to judge exactly. We can estimate the parameters with the phreg function.

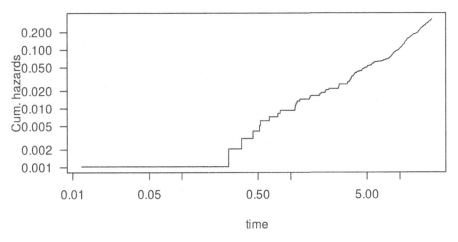

**FIGURE B.3** Weibull plot of male mortality data.

|           | log(scale) | log(shape) |
|-----------|-----------:|-----------:|
| Coef      | 3.785      | 0.332      |
| se(Coef)  | 0.066      | 0.058      |
| exp(Coef) | 44.015     | 1.393      |

The estimate of the shape parameter is 1.393, and it is significantly different from 1 $(p = 0)$, so we can firmly reject the hypothesis that data come from an exponential distribution.

### B.1.4  The Lognormal distribution

The lognormal distribution is connected to the normal distribution through the exponential function: If $X$ is normally distributed, then $Y = \exp(X)$ is lognormally distributed. Conversely, if $Y$ is lognormally distributed, then $X = \log(Y)$ is normally distributed.

The lognormal distribution has the interesting property that the hazard function is first increasing, then decreasing, in contrast to the Weibull distribution which only allows for monotone (increasing or decreasing) hazard functions.

The **R** functions are named (dpqrhH)lnorm.

### B.1.5   The Loglogistic distribution

The loglogistic distribution is very close to the lognormal, but has heavier tail to the right. Its advantage over the lognormal is that the hazard function has closed form. It is given by

$$h(t; (p, \lambda)) = \frac{\frac{p}{\lambda}(\frac{t}{\lambda})^{p-1}}{1 + (\frac{t}{\lambda})^p}, \quad t, p, \lambda > 0.$$

With shape $p = 2$ and scale $\lambda = 1$, its appearance is shown in Figure B.4.

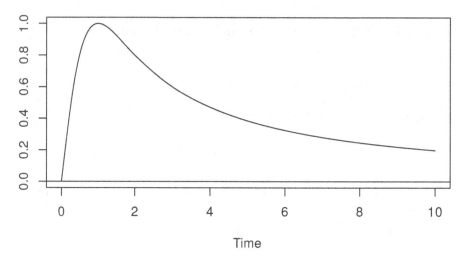

**FIGURE B.4** Loglogistic hazard function with shape 2 and scale 1.

It is produced by the code

```
x <- seq(0, 10, length = 1000)
plot(x, hllogis(x, shape = 2, scale = 1),
     type = "l", ylab = "", xlab = "Time")
abline(h = 0)
```

The **R** functions are named (dpqrhH)llogis.

**B.1.6 The Gompertz distribution**

The Gompertz distribution is useful for modeling old age mortality. The hazard function is exponentially increasing.

$$h(t; (p, \lambda)) = p \exp\left(\frac{t}{\lambda}\right), \quad t, p, \lambda > 0. \tag{B.2}$$

It was suggested by Gompertz (1825).

The **R** functions are named (dpqrhH)gamma.

**B.1.7 The Gompertz-Makeham distribution**

The Gompertz distribution was generalized by Makeham (1860). The generalization consists of adding a positive constant to the Gompertz hazard function,

$$h(t; (\alpha, p, \lambda)) = \alpha + p \exp\left(\frac{t}{\lambda}\right), \quad t, \alpha, p, \lambda > 0. \tag{B.3}$$

It is more difficult to work with than the other distributions described here. It is not one of the possible choices in the functions phreg or aftreg in **eha**.

The **R** functions are named (dpqrhH)makeham.

**B.1.8 The Gamma distribution**

The Gamma distribution is another generalization of the exponential distribution. It is popular in modeling shared frailty, see Hougaard (2000). It is not one of the possible distributions for the functions phreg and aftreg, and it is not considered further.

The **R** functions are named (dpqr)gamma.

## B.2 Proportional Hazards Models

Proportional hazards families of distributions are characterized by the property that if a distribution with hazard function $h(t), t > 0$ belongs to the family, so does all distributions with hazard functions $\Psi h(t), t > 0; \Psi > 0$. *Natural* proportional hazards families of distributions include the *Weibull, Exponential, Extreme value, Gompertz*, and *Piecewise constant hazard* families of distributions.

As an example, the proportional hazards Gompertz family of distributions is characterized by the hazard functions

$$h_{\Psi}(t) = \Psi \exp(\alpha t), \quad t > 0; \ \Psi > 0, \tag{B.4}$$

for each *fixed* $\alpha$, that is, by keeping $\alpha$ fixed and vary $\Psi$, one proportional hazards family of distributions is created. We note in passing that if $\alpha = 0$, then the resulting family of distributions is *exponential* with rate $\Psi$, and if $\alpha < 0$, then the distribution is degenerate (the probability of "eternal life" is positive). In the `phreg` function in **eha**, $\alpha$ is allowed to vary freely, if the parameterization `param = "rate"` is chosen. Note that it is *not* the default value (it may be in a near future).

Extended, three-parameter, versions of the Lognormal and Loglogistic are fit into the proportional hazards framework in the function `phreg`. It is simply done by adding a multiplicative positive constant to the hazard function as the third parameter. It is, at the time of this writing, experimental, so use with care.

## B.3 Accelerated Failure Time Models

Accelerated failure time families of distributions are defined by the property that if a distribution with survivor function $S(t), t > 0$

belongs to the family, so does any distribution with survivor function $S(\Psi t), t > 0; \Psi > 0$. Natural families with this property include the *Lognormal*, *Loglogistic*, *Weibull*, *Extreme value*, and *Exponential* families. With a small twist, also the *Gompertz* distribution can be (and is, in `aftreg`) included.

# C

## A Brief Introduction to **R**

There are already many excellent sources of information available for those who want to learn more about **R**. A visit to the **R** home page[1], is highly recommendable. When there, look under *Documentation*. The *Task Views* on the home page of CRAN[2] contains a section *Survival* worth examining.

When **R** is installed on a computer, it is easy to access the documentation that comes with the distribution; start **R** and read the help page `?help.start`. A good start for a beginner is also the book *Introductory Statistics with R* by (Dalgaard, 2008). The book on *S Programming* (Venables and Ripley, 2000) is recommended for the more advanced studies of the topic that its title implies.

That said, in this chapter we will only look at some aspects of working with **R**, aspects that are relevant to our topic.

### C.1   R in General

In this section we investigate how to work with **R** in general, without special reference to event history analysis, which will be treated separately in a later section in this chapter.

---

[1]`https://www.r-project.org`
[2]`https://cran.r-project.org`

DOI: 10.1201/9780429503764-C

### C.1.1  R objects

**R** is an object-oriented programming language, which have certain implications for terminology. So is for instance everything i **R** *objects*. We will not draw too much on this fact, but it may be useful to know.

### C.1.2  Expressions and assignments

Commands are either *expressions* or *assignments*. Commands are separated by newline (normally) or semi-colon. The hash mark (#) marks the rest of the line as a *comment*.

```
1 + exp(2.3)
```

```
[1] 10.97418
```

```
x <- 1 + exp(2.3)
x
```

```
[1] 10.97418
```

Note that the **R** *prompt* is >. If the previous expression is *incomplete*, it changes to +.

The first line above is an expression. It is evaluated as `return` is pressed, and the result is normally printed. The second line is an assignment, the result is stored in x and not printed. By typing x, its content is normally printed.

The assignment symbol is <-, which consists of two consecutive key strokes, "<" followed by "-". It is strongly recommended to enclose the assignment symbol (and any arithmetic operator) between spaces. It is *forbidden to separate* the two symbols by a space! You will then get comparison with something negative:

```
x <- 3
x < - 3
```

```
[1] FALSE
```

It ia also allowed, nowadays, to use = as an assignment symbol, but be aware of the fact that <- and = are not always exchangeable, for instance when giving values to arguments in function calls.

### C.1.3 Objects

Everything in **R** is an *object* (object-oriented programming). All objects have a *mode* and a *length*. Data objects have modes *numeric*, *complex, logical*, or *character*, and language objects have modes *function, call, expression, name*, etc. All data objects are *vectors*; there are *no scalars* in **R**.

Vectors contain elements of the same kind; a *list* can be thought of a vector where each element can be anything, even a list, and the elements can be completely different.

There are more properties of objects worth mentioning, but we save that for the situations where we need them.

### C.1.4 Vectors and matrices

There are five basic types of vectors: logical, integer, double, complex, and character. The function c (for *concatenate*) creates vectors:

```
dat <- c(2.0, 3.4, 5.6)
cols <- c("red", "green", "blue")
```

The elements of a vector may be *named*:

```
names(dat) <- c("Sam", "John", "Kate")
names(dat)
```

```
[1] "Sam"  "John" "Kate"
```

```
dat
```

```
 Sam John Kate
 2.0  3.4  5.6
```

```
dat["John"]
```

```
John
 3.4
```

Note the first two lines; names may either *display* (second row) or *replace* (first row).

The function *matrix* creates a matrix (surprise!):

```
Y <- matrix(c(1,2,3,4,5,6), nrow = 2, ncol = 3)
Y
```

```
     [,1] [,2] [,3]
[1,]    1    3    5
[2,]    2    4    6
```

```
dim(Y) <- c(3, 2)
Y
```

```
     [,1] [,2]
[1,]    1    4
[2,]    2    5
[3,]    3    6
```

```
as.vector(Y)
```

```
[1] 1 2 3 4 5 6
```

```
mode(Y)
```

```
[1] "numeric"
```

```
attributes(Y)
```

```
$dim
[1] 3 2
```

It is also possible to create vectors by the functions numeric, character, and logical.

```
x <- logical(7)
x
```

```
[1] FALSE FALSE FALSE FALSE FALSE FALSE FALSE
```

```
x <- numeric(7)
x
```

```
[1] 0 0 0 0 0 0 0
```

```
x <- character(7)
x
```

```
[1] "" "" "" "" "" "" ""
```

These functions are useful for *allocating* storage (memory), which later can be filled. This is better practice than letting a vector grow sequentially.

The functions `nrow` and `ncol` extracts the number of rows and columns, respectively, of a matrix.

An *array* is an extension of a matrix to more than two dimensions.

### C.1.5  Lists

Family records may be produced as a list:

```
fam <- list(FamilyID = 1, man = "John", wife = "Kate",
            children = c("Sam", "Bart"))
fam
```

```
$FamilyID
[1] 1

$man
[1] "John"
```

```
$wife
[1] "Kate"

$children
[1] "Sam"   "Bart"
```

```
fam$children
```

```
[1] "Sam"   "Bart"
```

## C.1.6   Data frames

A *data frame* is a special case of a *list*. It consists of variables of the same length, but not necessarily of the same type. Data frames are created by either `data.frame` or `read.table`, where the latter reads data from an ASCII file on disk.

```
dat <- data.frame(name = c("Homer", "Flanders", "Skinner"),
                  income = c(1000, 23000, 85000))
dat
```

```
      name income
1    Homer   1000
2 Flanders  23000
3  Skinner  85000
```

A data frame has *row names* (here: 1, 2, 3) and variable (column) names (here: `name`, `income`). The data frame is the normal object for data storage and statistical analysis in **R**.

## C.1.7   Factors

*Categorical* variables are conveniently stored as *factors* in **R**.

```
country <- factor(c("se", "no", "uk", "uk", "us"))
country
```

```
[1] se no uk uk us
Levels: no se uk us
```

```
print.default(country)
```

```
[1] 2 1 3 3 4
```

Factors are internally coded by integers (1, 2, ...). The levels are by default sorted into alphabetical order. Can be changed:

```
country <- factor(c("se", "no", "uk", "uk", "us"),
                  levels = c("uk", "us", "se", "no"))
country
```

```
[1] se no uk uk us
Levels: uk us se no
```

The first level often has a special meaning ("reference category") in statistical analyses in **R**.

## C.1.8   Operators

Arithmetic operations are performed on vectors, element by element. The common operators are + - * / ^, where ^ is exponentiation. For instance

```
x <- c(1, 2, 3)
x^2
```

```
[1] 1 4 9
```

```
y <- 2 * x
y / x
```

```
[1] 2 2 2
```

## C.1.9   Recycling

The last examples above were examples of *recycling*; If two vectors of different lengths occur in an expression, the shorter one is recycled (repeated) as many times as necessary. For instance

```
op <- options(warn = -1) # Turn off warnings.
x <- c(1, 2, 3)
y <- c(2, 3)
x / y
```

```
[1] 0.5000000 0.6666667 1.5000000
```

```
options(op) # Set warnings to original state.
```

so the actual operation performed is

```
c(1, 2, 3) / c(2, 3, 2) # Truncated to equal length
```

```
[1] 0.5000000 0.6666667 1.5000000
```

It is most common with the shorter vector being of length one, for instance, a scalar multiplication. If the length of the longer vector is not a multiple of the length of the shorter vector, a warning is issued: `longer   object length is not a multiple of shorter object length` (not shown above). This is often a sign of a mistake in the code (so do not turn off warnings!).

### C.1.10 Precedence

Multiplication and division is performed before addition and subtraction, as in pure mathematics. The (almost) full set of rules, from highest to lowest, are:

1. `$, [[`: Extraction of elements in a list.
2. `[`: Vector extraction.
3. `^`: Exponentiation.
4. `-`: Unary minus.
5. `:`: Sequence generation.
6. `%%, %/%, %*%`: Special operators.
7. `*, /`: & Multiplication and division.
8. `+, -`: Addition and subtraction.
9. `<, >, <=, >=, ==, !=`: Comparison.
10. `!`: Logical negation.
11. `&, |, &&, ||`: Logical operators.
12. $\sim$: Formula.

13.  <-: Assignment.

It is, however, highly recommended to use parentheses often rather than relying on (and remembering) those rules. Compare, for instance

```
1:3 + 1
```

```
[1] 2 3 4
```

```
1:(3 + 1)
```

```
[1] 1 2 3 4
```

For descriptions of specific operators, consult the help system.

## C.2 Some Standard R Functions

1.  round, floor, ceiling: Rounds to nearest integer, downward, and upward, respectively.

```
round(2.5) # Round to even
```

```
[1] 2
```

```
floor(2.5)
```

```
[1] 2
```

```
ceiling(2.5)
```

```
[1] 3
```

2.  `%/%` and `%%` for integer division and modulo reduction.

```
5 %/% 2
```

```
[1] 2
```

```
5 %% 2
```

```
[1] 1
```

3.  Mathematical functions: `abs`, `sign`, `log`, `exp`, etc.
4.  `sum`, `prod`, `cumsum`, `cumprod` for sums and products.
5.  `min`, `max`, `cummin`, `cummax` for extreme values.
6.  `pmin`, `pmax` parallel min and max.

```
x <- c(1, 2, 3)
y <- c(2, 1, 4)
max(x, y)
```

```
[1] 4
```

```
pmax(x, y)
```

```
[1] 2 2 4
```

7. `sort`, `order` for sorting and ordering. Especially useful is
   `order` for rearranging a data frame according to a specific
   ordering of one variable:

```
require(eha)
data(mort)
mt <- mort[order(mort$id, -mort$enter), ]
last <- mt[!duplicated(mt$id), ]
```

First, `mort` is sorted after `id`, and within `id` decreasingly after `enter`
(notice the minus sign in the formula!). Then all rows with duplicated `id` are removed, only the *first* appearance is kept. In this way
we get a data frame with exactly one row per individual, the row
that corresponds to the individual's last (in age or calendar time).
See also the next item!

8. `duplicated` and `unique` for marking duplicated elements in
   a vector and removing duplicates, respectively.

```
x <- c(1,2,1)
duplicated(x)
```

```
[1] FALSE FALSE  TRUE
```

```
unique(x)
```

```
[1] 1 2
```

```
x[!duplicated(x)]
```

```
[1] 1 2
```

### C.2.1 Sequences

The operator : is used to generate sequences of numbers.

```
1:5
```

```
[1] 1 2 3 4 5
```

```
5:1
```

```
[1] 5 4 3 2 1
```

```
-1:5
```

```
[1] -1 0 1 2 3 4 5
```

```
-(1:5)
```

```
[1] -1 -2 -3 -4 -5
```

```
-1:-5
```

```
[1] -1 -2 -3 -4 -5
```

Don't trust that you remember rules of precedence; use parentheses!

There is also the functions `seq` and `rep`, see the help pages!

## C.2.2 Logical expression

*Logical expressions* can take only two distinct values, TRUE and FALSE (note uppercase; **R** acknowledges the difference between lower case and upper case).

Logical vectors may be used in ordinary arithmetic, where they are *coerced* into numeric vectors with FALSE equal to zero and TRUE equal to one.

## C.2.3 Indexing

*Indexing* is a powerful feature in **R**. It comes in several flavors.

- **A logical vector** Must be of the same length as the vector it is applied to.

```
x <- c(-1:4)
x
```

```
[1] -1  0  1  2  3  4
```

```
x[x > 0]
```

```
[1] 1 2 3 4
```

- **Positive integers** Selects the corresponding values.
- **Negative integers** Removes the corresponding values.
- **Character strings** For named vectors; select the elements with the given names.
- **Empty** Selects all values. Often used to select rows/columns from a matrix,

```
x <- matrix(c(1,2,3,4), nrow = 2)
x
```

```
     [,1] [,2]
[1,]    1    3
[2,]    2    4
```

```
x[, 1]
```

```
[1] 1 2
```

```
x[, 1, drop = FALSE]
```

```
      [,1]
[1,]    1
[2,]    2
```

Note that a dimension is lost when one row/column is selected. This can be overridden by the argument drop = FALSE (*don't drop dimension(s)*).

### C.2.4  Vectors and matrices

A *matrix* is a vector with a dim attribute.

```
x <- 1:4
x
```

```
[1] 1 2 3 4
```

```
dim(x) <- c(2, 2)
dim(x)
```

```
[1] 2 2
```

```
x
```

```
      [,1] [,2]
[1,]    1    3
[2,]    2    4
```

```
t(x)
```

```
     [,1] [,2]
[1,]    1    2
[2,]    3    4
```

Note that matrices are stored *column-wise*. The function t gives the transpose of a matrix.

An example of matrix multiplication.

```
x <- matrix(c(1,2,3,4, 5, 6), nrow = 2)
x
```

```
     [,1] [,2] [,3]
[1,]    1    3    5
[2,]    2    4    6
```

```
y <- matrix(c(1,2,3,4, 5, 6), ncol = 2)
y
```

```
     [,1] [,2]
[1,]    1    4
[2,]    2    5
[3,]    3    6
```

```
y %*% x
```

```
        [,1] [,2] [,3]
[1,]      9   19   29
[2,]     12   26   40
[3,]     15   33   51
```

```
x %*% y
```

```
        [,1] [,2]
[1,]     22   49
[2,]     28   64
```

Dimensions must match according to ordinary mathematical rules.

### C.2.5  Conditional execution

Conditional execution uses the `if` statement (but see also `switch`). Conditional constructs are typically used *inside functions*.

```
x <- 3
if (x > 3) {
    cat("x is larger than 3\n")
}else{
    cat("x is smaller than or equal to 3\n")
}
```

```
x is smaller than or equal to 3
```

See also the function `cat` (and `print`).

## C.2.6   Loops

Loops are typically only used *inside functions.*

- A for loop makes an expression to to be iterated as a variable
  assumes all values in a sequence.

```
x <- 1
for (i in 1:4){
    x <- x + 1
    cat("i = ", i, ", x = ", x, "\n")
}
```

```
i =  1 , x =  2
i =  2 , x =  3
i =  3 , x =  4
i =  4 , x =  5
```

```
x
```

```
[1] 5
```

- The while loop iterates as long as a condition is TRUE.

```
done <- FALSE
i <- 0
while (!done & (i < 5)){
    i <- i + 1
    done <- (i > 3)
}
if (!done) cat("Did not converge\n")
```

- The `repeat` loop iterates indefinitely or until a `break` statement is reached.

```
i <- 0
done <- FALSE
repeat{
    i <- i + 1
    done <- (i > 3)
    if (done) break
}
i
```

```
[1] 4
```

## C.2.7 Vectorizing

In **R**, there are a family of functions that operates directly on vectors. See the documentation of `apply`, `tapply`, `lapply`, and `sapply`.

## C.3 Writing Functions

Functions are created by assignment using the keyword `function`.

```
fun <- function(x) exp(x)
fun(1)
```

```
[1] 2.718282
```

This function, `fun`, is the exponential function, obviously. Usually, the function body consists of several statements.

```
fun2 <- function(x, maxit){
    if (any(x <= 0)) error("Only valid for positive x")
    i <- 0
    sum <- 0
    for (i in 0:maxit){
        sum <- sum + x^i / gamma(i + 1)
    }
    sum
}
fun2(1, 5)
```

```
[1] 2.716667
```

```
fun2(1:5, 5)
```

```
[1]   2.716667   7.266667 18.400000 42.866667 91.416667
```

This function, fun2, uses the Taylor expansion of $\exp(x)$ for calculation. This is a very crude variant, and it is only good for positive $x$. It throws an error if fed by a negative value. Note that it is *vectorized*, i.e., the first argument may be a real-valued vector. Also note the function any.

In serious function writing, it is very inconvenient to define the function at the command prompt. It is much better to write it in a good editor and then source it into the **R** session as needed. If it will be used extensively in the future, the way to go is to write a package. This is covered later in these notes. Our favorite editor, and the one used in writing the examples in these notes, is emacs, which is free and available on all platforms. Another option is to use RStudio (RStudio Team, 2021).

Functions can be defined inside the body of another function. In this way it is hidden for other functions, which is important when considering scoping rules. This is a feature that is not available in Fortran 77 or C. Suppose that the function `inside` is defined inside the function `head`.

1. A call to `inside` will use the function defined in `head`, and not some function outside with the same name.
2. The function `inside` is not visible outside `head` and cannot be used there.

The *return value* of a function is the value of the last expression in the function body.

## C.3.1 Calling conventions

The arguments to a function are either *specified* or *unspecified* (or both). Unspecified arguments are shown as "..." in the function definition. See for instance the functions `c` and `min`. The "..." may be replaced by any number of arguments in a call to the function.

The *formal* arguments are those specified in the function definition and the *actual* arguments are those given in the actual call to the function. The rules by which formal and actual are *matched* are

1. Any actual argument given as `name = value`, where `name` exactly matches a formal argument, is matched.
2. Arguments specified by `name = value`, but no exact match are then considered. If a perfect *partial* match is found, it is accepted.
3. The remaining unnamed arguments are then matched *in sequence* (*positional* matching).
4. Remaining unmatched actual arguments are part of the ... argument if there is one. Otherwise an error occurs.
5. Formal arguments need not be matched.

## C.3.2    The argument "..."

The purpose of the ... argument is to pass arguments to a function call inside the actual function. Read *An Introduction to R* (available in the **R** help system) for further details.

## C.3.3    Writing functions

Below is a function that calculates the maximum likelihood estimate of the parameter in the exponential distribution, given a censored sample.

```
mlexp <- function(x, d){
    n <- length(x)
    ## Check indata:
    if (any(x <= 0)) stop("This function needs positive x")
    if (length(d) != n) stop("Lengths of x and d do not match")
    ## End check

    ## We adopt the rule: d == 0 for censored obs; 1 for 'events'.

    d <- as.numeric(d != 0)

    ## The density function is lambda^d * exp(-lambda * x), x > 0.
    ## But we prefer a parameterization where the parameter may vary freely.
    ## Set lambda = exp(alpha):
    ## The density is exp(d * alpha) * exp(-x * exp(alpha)).

    sumx <- sum(x)
    sumd <- sum(d)
    loglik <- function(alpha){
        sumd * alpha - sumx * exp(alpha)
    }
    ## We want the first derivative for optimization:
    dloglik <- function(alpha){
        sumd - sumx * exp(alpha)
    }
```

```
## Start value:
alpha <- 0
res <- optim(alpha, fn = loglik, gr = dloglik,
             method = "BFGS", control = list(fnscale = -1))
res
}
```

First note that the arguments of the function, x and d, do not have any default values. Both arguments must be matched in the actual call to the function. Second, note the check of sanity in the beginning of the function and the calls to stop in case of an error. For less severe mistakes it is possible to give a warning, which does not break the execution of the function.

### C.3.4 Lazy evaluation

When a function is called, the arguments are parsed, but no evaluated until it is needed in the function. Specifically, if the function does not use the argument, it is never evaluated, and the variable need not even exist. This is in contrast to rules in C and Fortran.

```
fu <- function(x, y) 2 * x
fu(3)
```

```
[1] 6
```

```
fu(3, y = zz)
```

```
[1] 6
```

```
exists("zz")
```

```
[1] TRUE
```

## C.3.5  Recursion

Functions are allowed to be *recursive*, i.e., a function in **R** is allowed to call itself. See *An Introduction to R* for details, but be warned: This technique may be extremely memory-intensive and slow. Use with caution, there is often a more direct way of attack.

## C.3.6  Vectorized functions

Most mathematical functions in **R** are *vectorized*, meaning that if the argument to the function is a vector, them the result is a vector of the same length. Some functions apply the recycling rule to vectors shorter than the longest:

```
sqrt(c(9, 4, 25))
```

```
[1] 3 2 5
```

```
pnorm(1, mean = 0, sd = c(1,2,3))
```

```
[1] 0.8413447 0.6914625 0.6305587
```

When writing own functions, the question of making it vectorized should always be considered. Often it is automatically vectorized, like

```
fun <- function(x) 3 * x - 2
fun(c(1,2,3))
```

```
[1] 1 4 7
```

but not always. Under some circumstances, it may be necessary to explicitly make the calculations separately for each element in a vector.

### C.3.7 Scoping rules

Within a function in **R**, objects are found in the following order:

1. It is a *locally* defined variable.
2. It is in the argument list.
3. It is defined in the environment where the function is defined.

Note that a function will eventually look for the variable in the work space. This is often *not* the intention!

**Tip:** Test functions in a *clean workspace*. For instance, start R with the flag `R --vanilla`.

## C.4 Standard Graphics

The workhorses in standard graphics handling in **R** are the functions `plot` and `par`. Read their respective documentations carefully; you can manipulate your plots almost without limitation!

See the code and Figure C.1 for the output.

```
source("plot2.R")
plot2
```

```
function ()
{
    x <- rnorm(100)
    y <- x + rnorm(100, sd = 0.1)
    oldpar <- par(mfrow = c(1, 2))
    on.exit(par(oldpar))
    plot(x, y, main = "Scatterplot of x and y")
    hist(y, main = "Histogram of y", probability = TRUE)
}
```

```
plot2()
```

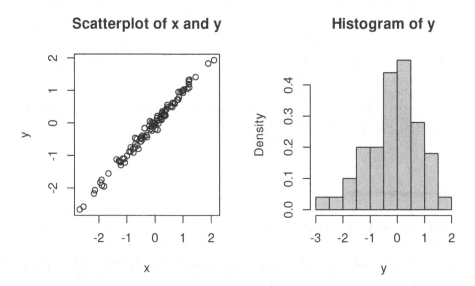

**FIGURE C.1** Output from the function plot2.

## C.5 Useful R Functions

### Matching

There are some useful functions in **R** doing *set operations*. For exact definitions of these functions, see the documentation.

- **match** `match(x, y)` returns a vector of the same length as x with the place of the first occurrence of each element in x in y. If there is mo match for a particular element in x, NA is returned (can be changed). Example

```
x <- c(1,3,2,5)
y <- c(3, 2, 4, 1, 3)
match(x, y)
```

```
[1]   4   1   2 NA
```

The `match` function is useful picking information in one data frame and put it in another (or the same). As an example, look at the data frame `oldmort`.

```
require(eha)
data(oldmort)
names(oldmort)
```

```
 [1] "id"          "enter"      "exit"       "event"     "birthdate"
 [6] "m.id"        "f.id"       "sex"        "civ"       "ses.50"
[11] "birthplace" "imr.birth"  "region"
```

Note the variables id (id of current record) and m.id (id of the mother to the current). Some of the mothers are also subjects in this file, and if we in an analysis want to to use mother's birthplace as a covariate, we can get it from her record (if present; there will be many missing mothers). This is accomplished with match; creating a new variable m.birthplace (mother's birthplace) is as simple as

```
oldmort$m.birthplace <- with(oldmort,
                    birthplace[match(m.id, id)])
```

The tricky part here is to get the order of the arguments to match right. Always check the result on a few records!

* **%in%** Simple version of match. Returns a logical vector of the same length as x.

```
x %in% y
```

```
[1]  TRUE  TRUE  TRUE FALSE
```

A simple and efficient way of selecting subsets of a certain kind, For instance, to select all cases with parity 1, 2, or 3, use

```
data(fert)
f13 <- fert[fert$parity %in% 1:3, ]
```

This is the same as

```
f13 <- fert[(fert$parity >= 1) & (fert$parity <= 3), ]
```

Note that the parentheses in the last line are strictly unnecessary, but their presence increases readability and avoids stupid mistakes about precedence between operators. Use parentheses often!

- **subset** Selects subsets of vectors, arrays, and data frames. See the help page.

- **tapply** Applies a function to all subsets of a data frame (it is more general than that; see the help page). Together with `match` it is useful for creating new summary variables for clusters and sticking the values to each individual.

If we for all rows in the `fert` data frame want to add the age at first birth for the corresponding mother, we can do this as follows.

```
data(fert)
indx <- tapply(fert$id, fert$id)
min.age <- tapply(fert$age, fert$id, min)
fert$min.age <- min.age[indx]
```

Check this for yourself! And read the help pages for `match` and `tapply` carefully.

## C.6 Help in R

Besides all documentation that can be found on CRAN[3], there is an extensive help system online in **R**. To start the general help system, type

```
> help.start()
```

---

[3]https://cran.r-project.org%7D

at the prompt (or, find a corresponding menu item). Then a help window opens up with access to several FAQs, and all the packages that are installed.

For immediate help on a specific function you know the name of, say coxreg, type

```
> ?coxreg
```

If you only need to see the syntax and input arguments, use the function args:

```
> args(coxreg)
```

## C.7  Functions for Survival Analysis

Here some useful functions in eha and survival are listed. In most cases the description is very sparse, and the reader is recommended to consult the corresponding help page when more detail is wanted.

- **aftreg** Fits accelerated failure time (AFT) models.
- **age.window** Makes a "horizontal cut" in the Lexis diagram, that is, it selects a subset of a data set based on age. The data frame must contain three variables that can be input to the surv function. The default names are enter, exit, and event. As an example,

```
library(eha)
mort6070 <- age.window(oldmort, c(60, 70))
```

limits the study of old age mortality to the (exact) ages between 60 and 70, i.e. from 60.000 up to and including 69.999 years of age.

For *rectangular* cuts in the Lexiz diagram, use both `age.window` and `cal.window` in succession. The order is unimportant.

- **cal.window** Makes a "vertical cut" in the Lexis diagram, that is, it selects a subset of a data set based on *calendar time*. As an example,

```
library(eha)
mort18601870 <- cal.window(oldmort, c(1860, 1870))
```

limits the study of old age mortality to the time period between 1860 and 1870, that is, from January 1, 1860 up to and including December 31, 1869.

For *rectangular* cuts in the Lexiz diagram, use both `age.window` and `cal.window` in succession. The order is unimportant.

- **cox.zph** Tests the proportionality assumption on fits from `coxreg` and `coxph`. This function is part of the survival package.

- **coxreg** Fits Cox proportional hazards model. It is a wrapper for `coxph` in case of simpler Cox regression models.

- **phreg** Fits parametric proportional hazards models.

- **pchreg** Piecewise constant proportional hazards regression

- **tpchreg** Piecewise constant proportional hazards regression for *tabular data.*

- **toTpch** Tabulating individual survival data.

- **risksets** Finds the members of each riskset at all event times.

## C.7.1 Checking the integrity of survival data

In data sources collected through reading and coding old printed handwritten sources, many opportunities for making and propagating errors occur. Therefore, it is important to have tools for detecting logical errors like people dying before they got married, of living at two places at the same time. In the sources used in this book, such errors occur in almost all data retrievals. It is important to say that this fact is not a sign of a bad job in the transfer of data from old sources to digital data files. The case is almost always that the errors are present in the original source, and in the registration process no corrections of "obvious" mistakes by the registrators are allowed; the digital files are supposed to be as close as possible to the original sources. These are the rules at the Demographic Database (DDB), Umeå University.

This means that the researcher has a responsibility to check her data for logical (and other) errors, and truthfully report how she handles these errors. In my own experience with data from the DDB, the relative frequency of errors varies between 1 and 5 per cent. In most cases it would not do much harm to simply delete erroneous records, but often the errors are "trivial", and it seems easy to guess what the truth should be.

### C.7.1.1 Old age mortality

As an example of the kind of errors discussed above, we take a look at the original version of oldmort in eha. For survival data in "interval" form (enter, exit], it is important that enter is smaller than exit. For individuals with more than one record, the intervals are not allowed to overlap, and at most one of them is allowed to end with an event. Note that we are talking about survival data, so there are no repeated events "by design".

The function check.surv in eha tries to do these kinds of checks and report back all individuals (id) that do not follow the rules.

```
load("olm.rda")
require(eha)
errs <- with(olm, check.surv(enter, exit, event, id))
errs
```

```
 [1] "785000980" "787000493" "790000498" "791001135" "792001121"
 [6] "794001225" "794001364" "794001398" "796000646" "797001175"
[11] "797001217" "798001203" "800001130" "801000743" "801001113"
[16] "801001210" "801001212" "802001136" "802001155" "802001202"
[21] "803000747" "804000717" "804001272" "804001354" "804001355"
[26] "804001532" "805001442" "806001143" "807000736" "807001214"
[31] "808000663" "810000704" "811000705" "812001454" "812001863"
[36] "813000613" "814000799" "815000885" "816000894" "817000949"
[41] "819001046"
```

```
no.err <- length(errs)
```

So, 41 individuals have bad records. Let us look at some of them.

```
badrecs <- olm[olm$id %in% errs, ]
dim(badrecs)
```

```
[1] 105  13
```

```
badrecs[badrecs$id == errs[1], ]
```

```
        id   enter   exit event birthdate       m.id       f.id
283 785000980 74.427 75.917 FALSE  1785.573 743000229 747000387
284 785000980 75.525 87.917 FALSE  1785.573 743000229 747000387
```

```
285 785000980 87.917 88.306   TRUE  1785.573 743000229 747000387
        sex     civ ses.50 birthplace imr.birth region
283 female married farmer      parish   12.4183  rural
284 female   widow farmer      parish   12.4183  rural
285 female   widow farmer      parish   12.4183  rural
```

The error here is that the second interval (75.526, 87.927] starts before the first interval has ended. We also see that this is a married woman who became a widow in the second record.

In this case I think that a simple correction saves the situation: Let the second interval begin where the first ends. The motivation for this choice is that the first interval apparently ends with the death of the woman's husband. Death dates are usually reliable, so I believe more in it than in other dates.

Furthermore, the exact choice here will have very little impact on the analyses we are interested in. The length of the overlap is less than half a year, and this person will eventually have the wrong civil status for at most that amount of time.

It turns out that all "bad" individuals are of this type. Therefore we "clean" the data set by running it through this function:

```
ger <- function(){
    require(eha)
    load("olm.rda")
    out <- check.surv(olm$enter, olm$exit, olm$event, olm$id)
    ko <- olm[olm$id %in% out, ]
    ## ko is all records belonging to "bad individuals"
    ## Sort ko, by id, then by interval start, then by event
    ko <- ko[order(ko$id, ko$enter, -ko$event), ]
    for (i in ko$id){ # Go thru all 'bad individuals'
        slup <- ko[ko$id == i, , drop = FALSE]
        ## Pick out their records
        n <- NROW(slup)
        if (n > 1){ # Change them
            for (j in (2:n)){
                slup$enter[j] <- slup$exit[j-1]
```

```
            }
            ## And put them back:
            ko[ko$id == i, ] <- slup
        }
    }
    ## Finally, put 'all bad' back into place:
    om <- olm
    om[om$id %in% out, ] <- ko
    ## Keep only records that start before they end:
    om <- om[om$enter < om$exit, ]
    om
}
```

Read the comments, lines (or part of lines) starting with #. The output of this function, om, is the present data file oldmort in eha, with the errors corrected as indicated above.

Running the function.

```
source("ger.R")
om <- ger()
left <- with(om, check.surv(enter, exit, event, id))
left
```

```
[1] "794001225"
```

Still one bad. Let us take a look at that individual.

```
om[om$id == left, ]
```

```
           id enter   exit event birthdate m.id f.id  sex   civ
1168 794001225 65.162 68.650 FALSE  1794.838   NA   NA male widow
1169 794001225 68.650 85.162 FALSE  1794.838   NA   NA male widow
1171 794001225 78.652 85.162 FALSE  1794.838   NA   NA male widow
```

```
        ses.50 birthplace imr.birth region
1168 farmer      parish  12.70903  rural
1169 farmer      region  12.70903  rural
1171 farmer      parish  12.70903  rural
```

It seems as if this error (second an third records overlap) should have been fixed by our function ger. However, there may have been a record removed, so we have to look at the original data file:

```
olm[olm$id == left, ]
```

```
            id     enter     exit event birthdate m.id f.id  sex    civ
1168 794001225 65.162 68.650 FALSE  1794.838   NA   NA male widow
1169 794001225 65.162 85.162 FALSE  1794.838   NA   NA male widow
1170 794001225 68.650 78.652 FALSE  1794.838   NA   NA male widow
1171 794001225 78.652 85.162 FALSE  1794.838   NA   NA male widow
        ses.50 birthplace imr.birth region
1168 farmer      parish  12.70903  rural
1169 farmer      region  12.70903  rural
1170 farmer      parish  12.70903  rural
1171 farmer      parish  12.70903  rural
```

Yes, after correction, the third record was removed. The matter can now be fixed through removing the last record in om[om\$id == left, ] :

```
who <- which(om$id == left)
who
```

```
[1] 1166 1167 1168
```

```
om <- om[-who[length(who)], ]
om[om$id == left, ]
```

```
             id    enter    exit event birthdate m.id f.id  sex    civ
1168 794001225 65.162 68.650 FALSE   1794.838   NA   NA male widow
1169 794001225 68.650 85.162 FALSE   1794.838   NA   NA male widow
      ses.50 birthplace imr.birth region
1168 farmer       parish  12.70903  rural
1169 farmer       region  12.70903  rural
```

And we are done.

---

## C.8 Reading Data into R

The first thing you have to do in a statistical analysis (in **R**) is to get data into the computing environment. Data may come from different kind of sources, and **R** has the capacity to read from many kinds of sources, like data files from other statistical programs, spreadsheet type (think *Excel*) data. This is normally done with the aid of the recommended package foreign (R Core Team, 2020), but the newer package haven (Wickham and Miller, 2020) is also useful.

If anything else fails, it is almost always possible to get data written in ASCII format (e.g., *.csv* files). These can be read into **R** with the function read.table, see the next section.

### C.8.1 Reading data from ASCII files

Ordinary text (ASCII) files in tabular form can be read into **R** with the function read.table. For instance, with a file looking this:

```
enter exit event
```

```
1    5    1
2    7    1
1    5    1
0    6    0
```

we read it into **R** with

```
mydata <- read.table("my.dat", header = TRUE)
```

Note the argument `header`. It can take two values, TRUE or FALSE, with FALSE being the default value. If `header =` TRUE, then the first row of the data file is interpreted as *variable names*.

There are a few optional arguments to `read.table` that are important to consider. The first is `dec`, which gives the character used as a decimal point. The default is ".", but in some locales "," (e.g., Swedish) is commonly used in output from other software than **R**. The second is `sep`, which gives the field separator character. The default value is "", which means"white space" (tabs, spaces, newlines, etc.). See the help page for all the possibilities, including the functions `read.csv`, `read.csv2`, etc., which are variations of `read.table` with other default values.

The result should always be checked after reading. Do that by printing a few rows

```
head(mydata)
```

```
  enter exit event
1     1    5     1
2     2    7     1
3     1    5     1
4     0    6     0
```

and/or by summarizing the columns

```
summary(mydata)
```

```
     enter            exit            event
Min.   :0.00    Min.   :5.00    Min.   :0.00
1st Qu.:0.75    1st Qu.:5.00    1st Qu.:0.75
Median :1.00    Median :5.50    Median :1.00
Mean   :1.00    Mean   :5.75    Mean   :0.75
3rd Qu.:1.25    3rd Qu.:6.25    3rd Qu.:1.00
Max.   :2.00    Max.   :7.00    Max.   :1.00
```

The str function is also helpful:

```
str(mydata)
```

```
'data.frame':   4 obs. of  3 variables:
 $ enter: int  1 2 1 0
 $ exit : int  5 7 5 6
 $ event: int  1 1 1 0
```

For each variable, its name and *type* is given, together with the first few values. In this example it is all the values, and we can also see that **R** interprets all the values as integer. This is fine, even though at least enter and exit probably are supposed to be real-valued. In fact, **R** is not so strict with numerical types (as opposed to *C* and *Fortran*). Frequent *coercing* takes place, and the *mode* numeric contains the types integer and double.

Now this tends to be confusing, but the bottom line is that you, as a user, need not worry at all. Except in one place: When you are writing an **R** function that calls compiled *C* or *Fortran* code, but that is far beyond the scope of this book. Interested readers are recommended to read Venables and Ripley (2000).

## C.8.2   Reading foreign data files

Data file from *Minitab*, *SAS*, *SPSS*, *Stata*, etc., can be read with a suitable function from the `foreign` package. Consult its help pages for more information. Another option is to use one of the `tidyverse` (Wickham et al., 2019) packages `haven`, `readxl`, and `DBI`. For more information, see relevant help pages.

# D

## Survival Packages in **R**

The basic package for survival analysis in **R** is the survival package (Therneau and Grambsch, 2000). It is one of the so-called *recommended packages* in **R**, which means that it is automatically installed when **R** itself is installed. You must however *load* it in a running **R** environment before you can use it.

There are a few other **R** packages devoted to survival and event history analysis. Besides eha, more or less the theme of this book, there are timereg and cmprsk. For a detailed explanation of how to use these packages, see their documentations in **R**.

### D.1  eha

The package eha is written and maintained by the author of this book. It has a long history as a stand-alone program (written in *Fortran, Turbo Pascal,* and *C* during different time periods) in the "pre **R**" era. When I was aware of the existence of the **R** environment in the mid-nineties, it was an easy decision to convert it into an **R** package.

Today the function coxreg in eha rests to a large part on the function coxph in the survival package, but it has some features of its own, notably

- **Discrete time Cox regression** With the option method = 'ml' a discrete-time Cox regression is performed with a discrete hazard atom at each observed event time. This is equivalent to a logistic regression with the *cloglog* link.

DOI: 10.1201/9780429503764-D

- **Sampling of risk sets** The *weird bootstrap* (Andersen et al., 1993) is implemented in `coxreg`. It is activated by setting the argument `boot` equal to the desired number of bootstrap replicates.
- **Time-dependent case weights**

Other features of the **eha** package are listed below.

- **Parametric proportional hazards models** The common parametric models in other packages are of the *AFT* type. While there is a function in `eha` for these models (`aftreg`), the function `phreg` fits proportional hazards parametric models. Especially worth mentioning is the implementation of the *piecewise constant hazard* (pch) model.
- **Lexis diagram cuts** With the aid of the two functions `age.window` and `cal.window` it is easy to make vertical and horizontal cuts in the Lexis diagram.
- **Tools for communal covariates** The main tool is the function `make.communal`, that takes an external time series (think *weather, economy, epidemics*, etc.) and turns it into a time-depedent covariate.

For a presentation of the most important functions in `eha`, see Appendix C and the on-line documentation.

---

## D.2 survival

The `survival` package is a recommended one, and it does not need a separate installation. It contains all the basic features that a package on survival analysis should have, and more. The main important functions are listed here:

- **coxph** This is the main function for Cox regression. It has a number of features, time dependent variables and strata, multiple events per subject, jackknife type variance estimators for clustered data, and frailty models. Allows left truncated data. Fast and reliable numerical algorithms.

- **survfit** Takes care of the presentation and "afterwork" of a fit to a proportional hazards model or a accelerated failure time model, including plotting and printing.
- **survreg** Fits parametric accelerated failure time models. Allows right and interval censoring, but not left truncation.
- **cox.zph** For testing of the proportionality assumption of fit from a call to coxph or coxreg.
- **aareg** Fits Aalen's additive hazards model (Aalen, 1989, 1993) to survival regression data.

## D.3 Other Packages

### D.3.1 coxme

The package coxme (Therneau, 2020) analyzes frailty models in Cox regression. Its author, Terry Therneau, is the author of the survival package, which also can fit frailty models. According to him, coxme is the prefered package for frailty models.

### D.3.2 timereg

The timereg package is developed by Martinussen and Scheike (2006). A key feature of the package (and the cited book) is extensions of the Cox model, especially models with time-varying effects of covariates. Aalen's additive hazards model is in focus. Resampling is frequently used for the calculation of $p$-values. The package has also been promoted for being able to analyze competing risks models (Scheike and Zhang, 2011).

### D.3.3   cmprsk

This is a competing risks package (Gray, 2020), based on work by
Gray (1988) and Fine and Gray (1999). It is used in the chapter
on competing risks models (Chapter 11).

# Bibliography

Aalen, O. (1978). Nonparametric inference for a family of counting processes. *Annals of Statistics*, 6:701–726.

Aalen, O. (1989). A linear regression model for the analysis of life times. *Statistics in Medicine*, 8:907–925.

Aalen, O. (1993). Further results on the non-parametric linear model in survival analysis. *Statistics in Medicine*, 12:1569–1588.

Aalen, O., Borgan, Ø., and Gjessing, H. (2008). *Survival and Event History Analysis: A Process Point of View*. Springer, New York.

Akaike, H. (1974). A new look at the statistical model identification. *IEEE Transactions on Automatic Control*, 19:716–723.

Allison, P. (1984). *Event History Analysis*. Sage Publications, Los Angeles.

Allison, P. (1995). *Survival Analysis using the SAS System: A Practical Guide*. SAS Institute Inc., Cary.

Andersen, P., Borgan, Ø., Gill, R., and Keiding, N. (1993). *Statistical Models Based on Counting Processes*. Springer-Verlag, Berlin.

Barbi, E., Lagona, F., Marsili, M., Vaupel, J. W., and Wachter, K. W. (2018). The plateau of human mortality: Demography of longevity pioneers. *Science*, 360:1459–1461.

Broström, G. (1987). The influence of mother's mortality on infant mortality: A case study in matched data survival analysis. *Scandinavian Journal of Statistics*, 14:113–123.

Broström, G. (2002). Cox regression: Ties without tears. *Communications in Statistics: Theory and Methods*, 31:285–297.

Broström, G. (2019). The semi-supercentenarian mortality plateau in Sweden. The 21st Nordic Demographic Symposium, Reykjavik, Iceland, June 13-15, 2019.

Broström, G. (2021). *eha: Event History Analysis.* R package version 2.9.0.

Broström, G. and Lindkvist, M. (2008). Partial partial likelihood. *Communications in Statistics: Simulation and Computation*, 37:679–686.

Broström, G. (2012). *Event History Analysis with R.* Chapman and Hall/CRC, Boca Raton, Florida. ISBN 978-1439831649.

Broström, G. and Bengtsson, T. (2011). Famines and mortality crises in 18th to 19th century southern Sweden. *Genus: Journal of Population Sciences*, 67:119–139.

Collett, D. (2003). *Modelling Survival Data in Medical Research.* Chapman & Hall/CRC, Second edition.

Cox, D. (1972). Regression models and life tables. *Journal of the Royal Statistical Society Series B (with discussion)*, 34:187–220.

Cox, D. (1975). Partial likelihood. *Biometrika*, 62:269–276.

Cox, D. and Oakes, D. (1984). *Analysis of Survival Data.* Chapman and Hall, London.

Dalgaard, P. (2008). *Introductory Statistics with R, Second Edition.* Springer, Berlin.

de Leeuw, J. (1992). Introduction to Akaike (1973) information theory and an extension of the maximum likelihood principle. In Kotz, S. and Johnson, N., editors, *Breakthroughs in Statistics I*, pages 599–609. Springer.

Efron, B. and Tibshirani, R. (1993). *An Introduction to the Bootstrap.* Chapman and Hall, New York.

Elandt-Johnson, R. and Johnson, N. (1999). *Survival Models and Data Analysis.* John Wiley & Sons, Inc., New York, Wiley Classics Library edition.

Fine, J. and Gray, R. (1999). A proportional hazards model for the subdistribution of a competing risk. *Journal of the American Statistical Association*, 94:496–509.

Gompertz, B. (1825). On the nature of the function expressive of the law of human mortality, and on a new mode of determining the value of life contingencies. *Philosophical Transactions of the Royal Society of London*, 115:513–585.

Granger, C. (1969). Investigating causal relations by econometric models and cross-spectral methods. *Econometrica*, 37:424–438.

Gray, B. (2020). *cmprsk: Subdistribution Analysis of Competing Risks.* R package version 2.2-10.

Gray, R. (1988). A class of K-sample tests for comparing the cumulative incidence of a competing risk. *Journal of tha American Statistical Association*, 94:496–509.

Groeneboom, P. and Wellner, J. (1992). *Nonparametric Maximum Likelihood Estimators for Interval Censoring and Deconvolution.* Birkhäuser, Boston.

Hall, P. and Wilson, S. (1991). Two guidelines for bootstrap hypothesis testing. *Biometrics*, 47:757–762.

Hauck, W. and Donner, A. (1977). Wald's test as applied to hyptheses in logit analysis. *Journal of the American Statistical Association*, 72:851–853.

Hernán, M., Brumback, B., and Robins, J. (2002). Estimating the causal effect of zidovudine on cd4 count with a marginal structural modelfor repeated measures. *Statistics in Medicine*, 21:1689–1709.

Hernán, M., Cole, S., Margolick, J., Cohen, M., and Robins, J. (2005). Structural accelerated failure time models for survival analysis in studies with time-varying treatments. *Pharmacoepidemiology and Drug Safety*, 14:477–491.

Hernán, M. A. and Robins, J. M. (2020). *Causal Inference: What If*. Chapman & Hall/CRC, London.

Hinkley, D. (1988). Bootstrap methods (with discussion). *J. Royal Statist. Soc. B*, 50:321–337.

Hougaard, P. (2000). *Analysis of Multivariate Survival Data*. Springer, Berlin.

Jackson, C. (2016). flexsurv: A platform for parametric survival modeling in R. *Journal of Statistical Software*, 70(8):1–33.

Johansen, S. (1983). An extension of Cox's regression model. *International Statistical Review*, 51:165–174.

Kalbfleisch, J. and Prentice, R. (1980). *The Statistical Analysis of Failure Time Data*. Wiley, Hoboken, N.J., First edition.

Kalbfleisch, J. and Prentice, R. (2002). *The Statistical Analysis of Failure Time Data*. Wiley, Hoboken, N.J., Second edition.

Klein, J. and Moeschberger, M. (2003). *Survival Analysis. Techniques for Censored and Truncated Data*. Springer-Verlag, New York.

Langholz, B. and Borgan, Ø. (1995). Counter-matching: A stratified nested case-control sampling method. *Biometrika*, 82:69–79.

Lauritzen, S. (1996). *Graphical models*. Oxford Statistical Science Series No. 17. Oxford University Press, Oxford, UK.

Lawless, J. (2003). *Statistical Models and Methods for Lifetime Data*. John Wiley & Sons, Hoboken, N.J., Second edition.

Lenart, A. and Vaupel, J. (2017). Questionable evidence for a limit to human lifespan. *Nature*, 546:E13–E14.

Makeham, W. (1860). On the law of mortality and the construction of annuity tables. *J. Inst. Actuaries and Assur. Mag.*, 8:301–310.

Martinussen, T. . and Scheike, T. (2006). *Dynamic Regression Models for Survival Data*. Springer/Verlag, New York.

Nelson, W. (1972). Theory and applications of hazard plotting for censored failure data. *Technometrics*, 14:945–965.

Parmar, M. and Machin, D. (1995). *Survival Analysis: A Practical Approach*. John Wiley & Sons, Chichester.

Pearl, J. (2000). *Causality: Models, Reasoning and Inference*. Cambridge University Press, New York.

Pinheiro, J. and Bates, D. (2000). *Mixed-Effects Models in S and S-Plus*. Springer-Verlag, New York.

R Core Team (2020). *foreign: Read Data Stored by 'Minitab', 'S', 'SAS', 'SPSS', 'Stata', 'Systat', 'Weka', 'dBase', ...* R package version 0.8-81.

R Core Team (2021). *R: A Language and Environment for Statistical Computing*. R Foundation for Statistical Computing, Vienna, Austria.

Robins, J. (1986). A new approach to causal inference in mortality studies with a sustained exposure period–application to control of the healthy worker survivor effect. *Mathematical Modeling*, 7:1393–1512.

Rootzén, H. and Zholud, D. (2017). Human life is unlimited–but short. *Extremes*, 20:713–728.

RStudio Team (2021). *RStudio: Integrated Development Environment for R*. RStudio, PBC, Boston, MA.

Rubin, D. (1974). Estimating causal effects of treatments in randomized and non-randomized studies. *Journal of Educational Psycology*, 66:688–701.

Scheike, T. and Zhang, M.-J. (2011). Analyzing competing risk data using the R timereg package. *Journal of Statistical Software*, 38(2):1–15.

Schweder, T. (1970). Composable Markov processes. *Journal of Applied Probability*, 7:400–410.

Silverman, B. (1986). *Density Estimation*. Chapman & Hall.

Therneau, T. (2020). *coxme: Mixed Effects Cox Models*. R package version 2.2-16.

Therneau, T. (2021). *survival: A package for survival analysis in R*. R package version 3.2-11.

Therneau, T. and Grambsch, P. (2000). *Modeling Survival Data: Extending the Cox Model*. Springer-Verlag, New York.

Vaupel, J., Manton, K., and Stallard, E. (1979). The impact of heterogeneity in individual frailty on the dynamics of mortality. *Demography*, 16:439–454.

Venables, W. and Ripley, B. (2000). *S Programming*. Springer-Verlag, New York.

Weibull, W. (1951). A statistical distribution function of wide applicability. *J. Appl. Mech.-Trans. ASME*, 18:293–297.

Wickham, H. (2021). *tidyr: Tidy Messy Data*. R package version 1.1.3.

Wickham, H., Averick, M., Bryan, J., Chang, W., McGowan, L. D., François, R., Grolemund, G., Hayes, A., Henry, L., Hester, J., Kuhn, M., Pedersen, T. L., Miller, E., Bache, S. M., Müller, K., Ooms, J., Robinson, D., Seidel, D. P., Spinu, V., Takahashi, K., Vaughan, D., Wilke, C., Woo, K., and Yutani, H. (2019). Welcome to the tidyverse. *Journal of Open Source Software*, 4(43):1686.

Wickham, H. and Miller, E. (2020). *haven: Import and Export 'SPSS', 'Stata' and 'SAS' Files*. R package version 2.3.1.

Wright, S. (1921). Correlation and causation. *Journal of Agricultural Research*, 20:557–585.

Xie, Y. (2015). *Dynamic Documents with R and knitr*. Chapman and Hall/CRC, Boca Raton, Florida, 2nd edition. ISBN 978-1498716963.

Xie, Y. (2016). *bookdown: Authoring Books and Technical Documents with R Markdown*. Chapman and Hall/CRC, Boca Raton, Florida. ISBN 978-1138700109.

Xie, Y. (2020). *bookdown: Authoring Books and Technical Documents with R Markdown*. R package version 0.21.

Xie, Y. (2021). *knitr: A General-Purpose Package for Dynamic Report Generation in R*. R package version 1.31.

Zeilinger, A. (2005). The message of the quantum. *Nature*, 438:743.

Zhu, H. (2021). *kableExtra: Construct Complex Table with 'kable' and Pipe Syntax*. R package version 1.3.4.

# *Index*

Printed in the United States
by Baker & Taylor Publisher Services